Leisure and Consumption

Also by Robert A. Stebbins

BETWEEN WORK AND LEISURE: The Common Ground of Two Separate Worlds

SERIOUS LEISURE: A Perspective for Our Time

PERSONAL DECISIONS IN THE PUBLIC SQUARE: Beyond Problem Solving into a Positive Sociology

Leisure and Consumption

Common Ground/Separate Worlds

Robert A. Stebbins
University of Calgary, Canada

First published 2009 by
PALGRAVE MACMILLAN

Palgrave Macmillan in the UK is an imprint of Macmillan Publishers Limited,
registered in England, company number 785998, of Houndmills, Basingstoke,
Hampshire RG21 6XS.

Palgrave Macmillan in the US is a division of St Martin's Press LLC,
175 Fifth Avenue, New York, NY 10010.

Palgrave Macmillan is the global academic imprint of the above companies
and has companies and representatives throughout the world.

Palgrave® and Macmillan® are registered trademarks in the United States,
the United Kingdom, Europe and other countries.

ISBN-13: 978–0–230–22022–5 hardback

This book is printed on paper suitable for recycling and made from fully
managed and sustained forest sources. Logging, pulping and manufacturing
processes are expected to conform to the environmental regulations of the
country of origin.

A catalogue record for this book is available from the British Library.

A catalog record for this book is available from the Library of Congress.

10   9   8   7   6   5   4   3   2   1
18 17 16 15 14 13 12 11 10 09

Printed and bound in Great Britain by
CPI Antony Rowe, Chippenham and Eastbourne

To Chris Rojek

Contents

List of Table and Figures   viii

Preface   ix

Acknowledgments   xii

1   The Nature of Leisure and Consumption   1

2   Conspicuous Consumption   30

3   Consumption and Leisure in Context   56

4   Phase One: Shopping   82

5   Phase Two: Consuming the Purchase   108

6   Organizing for Consumptive Leisure   134

Notes   160

Bibliography   163

Index   172

List of Table and Figures

Table

1.1 A Leisure-Based Theoretic Typology of Volunteers and
Volunteering   17

Figures

1.1 The Serious Leisure Perspective   14
6.1 Structural Complexity: From Tribes to Social Worlds   146

Preface

The social scientific study of consumption has been primarily concerned with a celebrated variety of this process known initially as mass consumption. Although this term is little used today, consumer studies remains as interested as ever in the modern, widespread propensity in the West to spend sometimes staggering amounts of money on popular goods and services. Moreover, most of the research and theory on such consumption has centered on its context – on its background social and personal conditions – with somewhat more work having been done on the first of these compared with the second. As a result the consumer's experience has been, in comparison, rather little studied, especially from the angle of leisure activity. Furthermore there is in the literature on mass consumption a tendency to say little about leisure other than to relate it generally to such consumption. This has created the impression among some observers that mass leisure experienced by way of mass consumption is the only leisure there is. This tendency has a long history, dating at least to the influential anthology assembled in 1958 by Eric Larrabee and Rolf Meyerson.

In the present book, however, I argue, following Ken Roberts and Jackie Kiewa, that in no way, can all of leisure can be equated with mass consumption, indeed with consumption of any kind. That is leisure and consumption are not always an identity. And this, even though as Daniel Cook writes, leisure 'is often seen as having been taken over by money and the money economy.' My goal here is to explore and clarify where consumption and taking leisure are separate processes, where they are similar if not the same, and in such overlap, what that looks like.

In their essence the two processes are clearly different. That is the end of consumption is to *have* something, to possess it, whereas the end of leisure is to *do* something, to engage in a positive activity. Nonetheless exceptions to this generalization exist, for there are times when consumption and leisure are so closely aligned as to make it impossible to distinguish the two in this way. Consider the hobbyist coin collector who travels abroad in search of a rare piece. A great amount and variety of knowledge is needed to succeed in this endeavor, such that the collector 'does' a lot (of leisure) to reach the point of buying the coin (to *have* it). It is the same with the person, for whom buying a new car is a joy (leisure), who does a great deal of research to determine, according to need, the best buy at the best price. As another example consider recreational gambling, where a player simultaneously places money and now, *having* gained the right to play, gambles by *doing* something like pull a lever or throw some dice. These cases exemplify the common ground shared by leisure and consumption.

Much of the time, however, leisure and consumption do indeed find themselves in obviously separate worlds. For one, consumption's relationship to leisure is, in substantial part, a practical matter. To be able to engage in a leisure activity, depending on its nature, the participant will have to buy at least one particular good or service. Furthermore the heart of the consumption-based leisure experience, which is found in 'doing' the core activity or activities, generally lies outside this practical expenditure (exceptions having just been noted). To analyze and accentuate the fact of this separateness, I treat of leisure-related consumption in two phases: shopping and consuming the purchase.

Looked on from another angle leisure and consumption occupy separate worlds, because for some kinds of leisure, monetary outlays are more or less unnecessary. That is, in a book claiming to explore the relationship between leisure and consumption, we are logically pushed to examine nonconsumptive leisure as well. There is far more of this kind of free-time activity than the mass leisure–consumption thesis would have us to believe, which is good news for low-income groups searching for something interesting but inexpensive to do when not faced with life's many obligations. For these reasons I go into some detail about nonconsumptive leisure at several points in this book.

At bottom leisure is not only doing something but also doing something positive. Still, its relationship to consumption is in fact more complicated than this. Thus throughout I consider consumptive behavior with reference to the three different forms of leisure – serious, casual, and project-based – and to their types and subtypes. The practical acts of buying, say, a cup of coffee over which to talk casually with friends and a new tennis racquet with which to play better one's favorite sport lead to two markedly different kinds of leisure experiences. The first, which is casual leisure, could also be classified today as mass leisure, whereas the second, which is serious leisure, certainly could not. More profoundly the second offers an avenue for self-fulfillment impossible to find in the first. In short, the leisure that follows from consumption, to the extent that consumption is even necessary for that leisure, varies immensely.

The scholarly literature on conspicuous consumption, mass leisure, mass consumption, McDonaldization, Dizneyization, the consumer society, and the like is, by contrast, predominantly negative, in the sense that its contributors are, in diverse ways, critical in their understanding of the relationship between consumption and leisure (a few exceptions exist as well). Yet extensive negativeness about something as positive as leisure should alert us to a problem. Leisure is fundamentally positive, and when searching for positiveness in their lives, many people turn to it. Positiveness is what people get when they organize their lives such that those lives become, in combination, substantially rewarding, satisfying, and fulfilling. Much of the time in contemporary Western society people must buy a good or service to engage in a positive activity, which is why consumption and leisure make natural partners. The question is, then, how can something as positive as leisure be seen by so many thinkers largely, if not wholly, in negative terms?

Part of the answer to this question is that the critics lack a sufficiently profound understanding of leisure. Part of the answer lies with the fault-finding proclivity of much of modern social science, about which I say more in my book *Personal Decisions in the Public Square*. So the present book has some important work to do: to correct the supposition that the partnership between leisure and consumption is wholly or even dominantly negative. The evidence from leisure studies indicates that this supposition, stated as it is in the extreme, is empirically untenable.

Acknowledgments

I want to thank Philippa Grand for taking the initiative to track me down and for encouraging me to submit the proposal that eventually became this book. She saw a monograph in the paper I presented in 2007 at the annual conference of the Leisure Studies Association (LSA), a subtle feature of my own work that I had missed. It was not until I got into writing that proposal that I saw the book in the LSA paper that she saw. She then had the perspicacity to line up a couple of fine anonymous reviews, from which my contribution to this work has benefited mightily.

Once writing began the editorial ball was passed to Olivia Middleton, who efficiently answered my queries about matters like style and deadlines. Cherline Daniel oversaw the copy-editing, proof-reading, and indexing phase of production. As with the first two, her work was cheerful, accurate, and punctual. In short, preparation of this book has been blessed with a superb editorial team reminding me once again how important this complicated, but unheralded, function is in publishing.

1
The Nature of Leisure and Consumption

Consumption studies, which describes and explains consumers, their consumptive behavior, and the creation, distribution, and purchase of the goods and services they consume, is a decidedly interdisciplinary field. Clarke, Doel, and Housiaux (2003, pp. 3–20) observe that, scientifically, consumption has been approached in three different ways, resulting in three identifiable bodies of literature. Economics and marketing look at it from the financial angle of the production and distribution of wealth. Anthropology, sociology, and cultural studies examine the social and cultural implications of consumption. History and geography explore consumptive practices across time and space. Such concerns as politics, ethics, and aesthetics, these authors observe, cut across these three bodies.

The early stages of consumption studies revolved mainly around economics, primarily the measurement of production and product consumption, and around marketing as expressed in research on consumer motivation to learn what customers want and what sorts of products interest them. Soon, however, abstract considerations emerged in the fields of sociology and cultural studies. Here thinkers began pondering such questions as the mass appeal of products, the consequences of popular consumptive patterns for the larger society, and the gendered basis of consumption, often taking a critical, if not moralistic, stand on their subject. Changes in consumptive practices over the years have also interested historians, while anthropologists have studied consumption cross-culturally, often in preliterate societies.

Yet consumption studies is not as interdisciplinary as it might be, and in this book I will argue, as it should be. As evidence note the observation by Zukin and Maguire (2004, p. 173) that sociologists in the United States 'have generally ignored the topic of consumption,' while those in

Europe have tended to line up more directly under the banners of such allied fields as cultural studies, feminist scholarship, and social history. Taking the sociological standpoint, Zukin and Maguire note that consumption is a vast subject, sprawling across all social institutions, these collective representations forming a distinctive prism through which sociologists have tended to view society.

From the institutional angle a good deal of consumption of goods and services may be said to be related to work, such as buying clothing and paying for transportation, but much of it, too, may be seen as familial (e.g., purchasing groceries, tickets to an amusement park). Quite clearly consumption has an economic function, while a lot of it is related to education (books, computers, field trips, etc.) and, not inconsequentially, to religion (e.g., religiously motivated arrangements related to baptism, marriage, and death). And then there is consumption carried out in the name of art, sport, science, and entertainment (some of which is also familial or educational). And, lest we forget, politics and government are part of the interdisciplinary formula, as seen in, for instance, regulations limiting store and bar hours, safety codes for various products, and taxes on goods and services. No wonder some sociologists have had difficulty getting their own intellectual grip on so diverse, amorphous, and widespread a field (as opposed to a grip on an established interdiscipline or an allied field like cultural studies).

Nonetheless the gap in the interdisciplinarity of consumption studies on which this book is centered is different. Here we look into the failure of specialists in this domain to systematically consider leisure in their explanations of consumers and their consumptive behavior. As with consumption leisure may be analyzed and understood through the lens of all the institutions just mentioned, even work (e.g., Stebbins, 2004a; in press). Yet leisure in these institutions is by no means always consumptive. In fact, leisure and consumption have a very complex relationship with each other. I argue in the following pages, after Ken Roberts (1999, p. 179) and Jackie Kiewa (2003, p. 80), that in no way can all of leisure can be equated with consumption, even mass consumption.

In other words, leisure and consumption are not always an identity. But Daniel Cook, for example, fails to see it this way: 'We don't live near or beside consumer society, but within it. Consequently we don't seek, experience, make or find leisure and recreation anywhere else' (Cook, 2006, p. 313). McDonald, Wearing, and Ponting (2007, p. 495) hold that leisure 'has become an escape from the pressures of the competitive individualized labour market through the process of therapeutic consumption. The importance placed upon the acquisition and consumption of commodities has resulted in fetishism ... over-consumption, luxury fever ... .' That is, it is common in scholarly circles to view leisure as little more than purchase of a good or service. My over-all goal, as we move through the chapters of this book, is to clarify and explain where consumption and taking leisure are separate processes, where they are similar if not the same, and in such overlap, what that looks like. This clarification and explanation is missing in the literatures on consumption and leisure.

The aim of the present chapter is to lay down the conceptual frame-work needed to work toward the promised clarification and explanation. This is accomplished in two major sections: the nature of consumption, which is the basic process in consumption studies, and the nature of leisure. In subsequent chapters these two approaches will perform on the same stage, as I try to show where they are separate actors, where they are similar if not the same, and in such closeness of character, what that looks like.

The nature of consumption

From the standpoint of the buyer, purchasing something is often far more complicated than the simple act of exchanging of money for a good or service. Consider the common, mundane act of buying some AA batteries. For one person this act is an annoyance, because those power-ing the travel alarm clock have gone dead just before departure on a long business trip. For another the new batteries are acquired with enthu-siasm, for they will enable the purchaser to use a new GPS-operated personal navigator, long awaited to facilitate route finding on backpack-ing trips. A third person buys some of these batteries while shopping for something else, on the realization, triggered by product display, that the supply at home is running low and should therefore be replenished. This buyer basks, however momentarily, in the self-congratulatory light of being personally prudent and foresightful.

Only the second purchase is clearly related to leisure. The first is in service of a work-related obligation. The third is neutral, in the sense that keeping a supply of AA batteries on hand can conveniently serve any future want or interest, however defined by the user. Many other purchases of, for instance, food, clothing, and petrol could be shown to generate the same or similar patterns of allied sentiments.

Consumption, says Russell Belk (2007, p. 737), 'consists of activities potentially leading to and actually following from the acquisition of a good or service by those engaging in such activities.' We are dealing here with *monetary acquisition*, defined in this book as either buying or renting with money a good or service. Bartering, borrowing, stealing, begging, and other forms of nonmonetary acquisition are deliberately excluded from this definition and hence from these pages. Each of these forms is sufficiently different sociologically to warrant separate treatment, lengthy undertakings that would take us too far afield from the goals of this book. Moreover consumption through monetary acquisition is intentional. As such receiving gifts falls beyond the purview of this work, because giving the gift roots in the intention of the giver rather than the receiver. Nevertheless a giver of a gift may present the gift as, in whole or in part, a conspicuous expression of consumptive power, a process examined in Chapter 2.

More particularly the social scientific study of consumption has been primarily concerned with a celebrated variety of this process, namely mass consumption, or the wide spread acquisition of popular goods and services (with these two sometimes referred to as mass culture). Further there is in the literature on mass consumption a tendency to say little about leisure other than to relate it to such consumption, thereby creating the impression that mass leisure is the only leisure that exists. As noted in the next section this tendency has a long history.

We will consider both parts of Belk's definition – 'activities potentially leading to and actually following from the acquisition of a good or service by those engaging in such activities.' The second part will be taken up in Chapter 5 as the second phase of consumption. Discussion of the first phase – activities potentially leading to acquisition of a good or service – will for the most part be limited to Chapter 4. Phase One of consumption, compared with Phase Two, has been well examined, both conceptually and empirically, under the banner of 'shopping' (e.g., Prus and Dawson, 1991; Bowlby, 1997; Falk and Campbell, 1997; Stebbins, 2006). Whereas what follows after the act of acquiring a good or service, the second phase of consumption, has, from the angle of leisure, been comparatively unexplored. My underlying model is that the act of monetary acquisition (i.e., purchasing or renting) of a good or service stands between these two phases, demarcating them, in the old days often and figuratively by the melodic ring of a cash register, but today, more often than not, by the muted hiss of a bank card being swiped. Note that the informal, or nonmonetary, borrowing of something, since no commodity is traded in this process, is not regarded here as consumptive acquisition. This two-phase model is elaborated further at the beginning of Chapter 4.

Mass consumption

This locution, never terribly well defined even in intellectual circles, has two, often overlapping, meanings. One refers to the consumptive practices of the masses, usually meaning the so-called numerically preponderant lower socioeconomic levels of society such as clerks, blue-collar workers, and manual labors. The other meaning centers on articles bought by large numbers (masses) of people, whose individual identities may span class boundaries, showing allegiance instead to other demographic dimensions, prominent among them, age, sex, and leisure activity, and to customization of leisure interests (see next section). Teenage popular music, age-graded clothing fashions, and plasma/LCD television sets exemplify the demographic understanding of mass consumption.

Herbert Gans has provided a useful overview of the mass leisure/mass culture critique that some analysts made of its perceived socially and psychologically harmful effects. This critique, more than any other force, set the stage for the rise of the contemporary view that leisure is largely, if not entirely, a matter of consumption, as an identity with this process. Gans (1974, pp. 4–5) starts by noting that such an assessment is endemic to urban industrial society, which in the 18th century saw work time and free time become separate periods in everyday life. Popular literature was the object of criticism during that era, which shifted in the 19th century to a focus on the 'iniquities' of alcohol and illicit sex and then later on the deleterious effects of sedentary spectator sport and televised entertainment. These practices, it was argued, lead to boredom, unhappiness, possibly even social chaos.

> With respect to leisure facilities in the nineteenth century, the critique emphasized the allegedly harmful effects of the music hall, the tavern, and the brothel; from about the 1920s to the 1950s, it focused on the movies, comic books, radio, and spectator sports. During the affluence of the 1950s, it expanded its concern to mass consumption in general, and to the suburban life-style in particular, but in the 1960s it narrowed again, centering now mainly on the negative effects of television viewing.
>
> (Gans, 1974, p. 4)

Though the object of criticism changed over this period, the theme remained much the same: popular leisure, as defined by the critics, is dangerous for both individual and society. Two influential anthologies assembled by Larrabee and Meyerson (1958) on mass leisure and by Rosenberg and White (1957) on mass culture gave intellectual expression to this stance.

Gans (1974, p. 5) goes on to note that, by the late 1960s, the mass leisure critique had come to an ideological fork in the road. Figuratively speaking, some critics took the leftish route, railing against what they saw as a youth culture founded on political radicalism, hedonism, mysticism, and nihilism. Of interest in the present book is the hedonic, consumerist component of this culture. Herbert Marcuse, a philosophical and sociological Marxist, became a key figure in criticizing this tendency, arguing in *One-Dimensional Man* (Marcuse, 1964) that mass culture is a potent repressive force. Disguised as affluence and liberty, technological rationality has become the major instrument of social domination. Better living supported by more and more gadgets is increasing alienation, he maintained. Humanity is willfully accepting its own domination by technology. 'People recognize themselves in their automobile, hi-fi set, split-level home, kitchen equipment.... There is only one dimension, and it is everywhere and in all forms' (Marcuse, 1964, pp. 9, 11).

Up the 'rightish' route went those who regarded the mass culture critique as dead, because high culture had by then been embraced at the popular cultural level. That is they saw cultural choice as no longer determined by class position. This proposition was most visibly championed by Nathan Glazer (1971) and Daniel Bell (1970). Gans (1974, p. 6), however, found this assessment insufficiently nuanced. He agreed that high cultural standards had been adopted outside the upper class, most particularly by the upper-middle class, but it was still possible to differentiate high and popular culture. The former only had a broader base than it previously had.

The mass leisure/mass society debate has, since the early 1970s, given way to a number of more specialized polemics such as the nature and impact of popular culture and the cultural industries. Of concern in this book is the continuation of the putative identity of leisure and consumption. Linking the two this way is common practice in discussions of the consumer society.

The consumer society

The belief that we presently live in a consumer society, flows logically from the belief that many kinds of leisure and culture have mass appeal. This reasoning assumes, however, that attractive leisure and culture are purchased (on a mass basis). It also assumes that most people in the consumer society have plenty of money to spend on purchases of hedonic good and services and that they therefore buy correspondingly. So central is consumption in this argument that, according to Zygmunt Bauman (2003 p. 26) the individual member 'needs to be a consumer first, before one can think of becoming anything in particular.' In other words, we are known by what we consume and, to be sure, are also socially ranked by that same criterion. This applies both in leisure and outside of it.

Chas Critcher (2006, p. 281) holds that commercialized leisure is an essential feature of the consumer society; leisure is something bought in the marketplace. Citing Celia Lury (1996) he lists several characteristics of the consumer society:

- Mass availability of consumer goods.
- Penetration of the market into almost all spheres of life.
- Use of advertising and promotion to attract consumers.
- Popularity of shopping in all its forms.
- Importance of selling and buying style, perhaps even lifestyle (Critcher, 2006, p. 281).

These characteristics are typical of the economy as a whole, though with emphasis on sport and leisure as commercial enterprise. These two, Critcher observes, also serve as vehicles for marketing goods and services in other spheres of life.

Critcher goes on to note that the rise of the consumer society has reduced the importance of social class as a correlate of leisure activities. The contemporary process of leisure customization helps explain this tendency.

Consumption and leisure customization

Mass consumption is still with us in the 21st century, though in the words of Geoffrey Godbey, it is now noticeably more 'customized' than earlier. That is mass consumption and mass leisure continue as before, exemplified in eating for pleasure at MacDonald's, watching popular films, listening to popular music, and attending games of the local professional team. Western nations are, among other things, consumer societies, and, as argued in this book, a significant part of consumption there may be qualified as leisure or leisure-related activities. Moreover we shall see in Chapter 3 that the social scientific critique of mass consumption continues apace.

Meanwhile overlaying the mass consumption of today is a trend toward shaping the leisure consumed to suit particular categories of consumers (Godbey, 2004). Mass leisure has always had a clear sense of equality about it – it is for everyone. But now, though mass leisure is still enjoyed, another kind of leisure is growing alongside it. This leisure is 'appropriate'; it is customized by or for special categories of society. And the emergence of these categories reflects some of the recent transformations to postmodern society. Godbey explains how these groups emerge in the wake of today's brisk pace of social change.

One, he notes the explosive population growth in the West. It is fueled substantially by international migration, which brings different sets of leisure-related customs. The efflorescence of 'ethnic' restaurants and grocery stores is evidence of this, as is the demand for special recreational facilities for Muslim women (VandeSchoot, 2005). Films shown in cinemas catering to particular nationalities constitute another sign of customization along lines of immigrant tastes. How many immigrants from Germany, Italy, Japan, or Korea, for example, buy cars in their new country that were made in the old country, cars used for leisure and for work and other obligations? Now we find video and CD shops catering to certain ethnic tastes in film and music. In the area of illegal deviant leisure, roosters are raised and sold for cockfighting in North America by immigrants from Asia and Latin America where this practice is legal (Pynn, 2008).

Two, composition of the family has changed considerably, with many more single-parent units, gay and lesbian unions, and multiethnic and multiracial groups. In most Western countries today a number of hotels operate according to a 'gay friendly' policy. Three, the world's population is aging. This has been paralleled by an increase in leisure programs and services for the elderly exemplified in their own exercise programs, musical performances, dance sessions, and local tours. Crafts are popular with older people, which leads some of them to take courses in this area and many of them to patronize stores supplying relevant material and tools to work with.

Four, work is, for many people, different now from what it used to be. Today the typical worker is always learning something new. Part-time jobs are now more common, as is work at home. Some seniors continue working part-time. Writing on new leisure I observed (Stebbins, 2009a) that part-time employment opens up workers to free-time opportunities as diverse as making up languages and practicing money slavery (where males make monetary payments to women in exchange for being humiliated and degraded over the Internet). And for those in the workforce suffering from a time famine, they may now play with dispatch the new 'express' board games.

Five, Godbey holds that today's economy is, in many respects, an experience economy. Evidence for this trend is ubiquitous. Thus, the rich may have the expensive, though highly unusual, experience of spaceflight, as provided by the Russian Space Agency. Nowadays people pay to ride on a luge run or a zip line, be pulled in a dog sled, go rafting on a raging river, eat at a fine restaurant, enjoy a spa, ride a helicopter to a mountain top, and similar thrills. All this for the distinctive experience each can bring.

Six, religion is also becoming more diverse in the West, thereby forcing customization of leisure services to accommodate the diverse free-time interests that are commensurate with religious principles. The example above about special recreational services for Muslin women fits here, too, as do some of the ethnic restaurants and grocery stores, their ethnicity in this instance being religious (e.g., those vending vegetarian cuisine; kosher food; special meat, fish, or foul for sacred holidays). Religion and custom also lead people to different leisure while celebrating their own special days. Ramadan, Ukrainian Christmas, and the Chinese New Year illustrate these practices.

The 'galactic city' (Lewis, 1995) is an increasingly common urban phenomenon. This kind of community – it spreads over considerable territory toward the outer edge of a larger metropolitan area while having no clear center – encourages use of the automobile and involvement in local leisure. This situation helps explain the popularity of, for example, neighborhood conversation cafés. These casual leisure, sociable conversations operate, often fortnightly, on a no-charge basis, and are held in a public setting, usually a local café. Anyone may participate, which is done by speaking in turn on a mutually agreed-upon subject. This session is followed by open dialogue. A skilled host leads the session.

The nature of leisure

Before turning to the serious leisure perspective – the vantage point we will use to examine the relationship of leisure and consumption – we must first look at three basic concepts that not only undergird this framework but also help clarify and explain that relationship. These concepts are leisure, general activity (and role), and core activity. They are also central to positive sociology, defined as the study of what people do to organize their lives such that their lives become, in combination, substantially rewarding, satisfying, and fulfilling (Stebbins, 2009b, pp. 1–2). Because leisure is inherently positive and positiveness through leisure is a fundamental line of inquiry for positive sociology, reference to this condition and the branch of knowledge devoted to studying and promoting it will surface at various points in this book.

What is leisure?

Scientifically speaking, leisure is uncoerced activity undertaken during free time. Uncoerced activity is positive activity that, using their abilities and resources, people both want to do and can do at either a personally satisfying or a deeper fulfilling level (Stebbins, 2005a, 2007a, pp. 4–5). Let us be clear at the outset about the place of boredom in everyday life. This state of mind is more complex than sometimes acknowledged. For boredom occurring in free time is a *coerced* state; it is not something bored people *want* to experience. Therefore it is not leisure; it is not a positive experience, as just defined. In fact any activity may be boring, be it free-time, work, or nonwork obligation. In these circumstances the desired end and the means to it are uninspiring. In free time the boring activity (e.g., hanging out on a street corner with nothing else to do, watching uninteresting television) is commonly the only activity seen by participants as available to them.

Uncoerced, people in leisure believe they are doing something they are not pushed to do, something they are not disagreeably obliged to do. In this definition emphasis is *ipso facto* on the acting individual and the play of human agency. This in no way denies that there may be things people want to do but cannot do because of any number of constraints on choice, because of limiting social and personal conditions; for example, aptitude, ability, socialized leisure tastes, knowledge of available activities, and accessibility of activities. In other words, when using this definition of leisure, whose central ingredient is lack of coercion, we must be sure to understand leisure activities in relation to their larger personal, structural, cultural, and historical background. And it follows that leisure is not really freely chosen, as some observers have claimed (e.g., Parker, 1983, pp. 8–9; Kelly, 1990, p. 7), since choice of activity is significantly shaped by this background.

Activity and role

An *activity* is a type of pursuit, wherein participants in it mentally or physically (often both) think or do something, motivated by the hope of achieving a desired end. Life is filled with activities, both pleasant and unpleasant: sleeping, mowing the lawn, taking the train to work, having a tooth filled, eating lunch, playing tennis matches, running a meeting, and on and on. Activities, as this list illustrates, may be categorized according to whether they fall within the domain of work, leisure, or nonwork obligation.[1] They are, furthermore, general. In some instances they refer to the behavioral side of recognizable roles, for example commuter, tennis player, and chair of a meeting. In others we may recognize the activity but not conceive of it so formally as a role, exemplified in someone sleeping, mowing a lawn, or eating lunch (not as patron in a restaurant).

The concept of activity is an abstraction, and as such, is broader than that of role. In other words, roles are associated with particular statuses, or positions, in society, whereas with activities, some are statuses and others are not. For instance, sleeper is not a status, even if sleeping is an activity. It is likewise with lawn mower (person). Sociologists, anthropologists, and psychologists tend to see social relations in terms of roles, and as a result, overlook activities whether aligned with a role or not. Meanwhile certain important parts of life consist of engaging in activities not recognized as roles. Where would many of us be were we unable to routinely sleep or eat lunch?

Moreover another dimension separates role and activity, namely, that of statics and dynamics. Roles are static whereas activities are dynamic.[2] Roles, classically conceived of, are relatively inactive expectations for behavior, whereas in activities, people are actually behaving, mentally or physically thinking or doing things to achieve certain ends. This dynamic quality provides a powerful explanatory link between an activity and a person's motivation to participate in it. Nevertheless the idea of role *is* useful in positive sociology, since participants do encounter role expectations in certain activities (e.g., those in sport, work, volunteering). Although the concept of activity does not include these expectations, in its dynamism, it can, much more effectively than role, account for invention and human agency.

This definition of activity gets further refined in the concept of *core activity*: a distinctive set of interrelated actions or steps that must be followed to achieve the outcome or product that the participant seeks. As with general activities, core activities are pursued in work, leisure, and nonwork obligation. Consider some examples in serious leisure: a core activity of alpine skiing is descending snow-covered slopes, in cabinet making it is shaping and finishing wood, and volunteer fire fighting is putting out blazes and rescuing people from them. In each case the participant takes several interrelated steps to successfully ski down hill, make a cabinet, or rescue someone. In casual leisure core activities, which are much less complex than in serious leisure, are exemplified in the actions required to hold sociable conversations with friends, savor beautiful scenery, and offer simple volunteer services (e.g., handing out leaflets, directing traffic in a theater parking lot, clearing snow off the neighborhood hockey rink). Work-related core activities are seen in, for instance, the actions of a surgeon during an operation or the improvisations on a melody by a jazz clarinetist. The core activity in mowing a lawn (nonwork obligation) is pushing or riding the mower. Executing an attractive core activity and its component steps and actions is a main feature drawing participants to the general activity encompassing it, because this core directly enables them to reach a cherished goal. It is the opposite for disagreeable core activities. In short, the core activity has motivational value of its own, even if more strongly held for some activities than others and even if some activities are disagreeable but still have to be done.

Core activities can be classified as simple or complex, the two concepts finding their place at opposite poles of a continuum. The location of a core activity on this continuum partially explains its appeal or lack thereof. Most casual leisure is comprised of a set of simple core activities. Here *homo otiosus* (leisure man) need only turn on the television set, observe the scenery, drink the glass of wine (no oenophile is he), or gossip about someone. Complexity in casual leisure increases slightly when playing a board game using dice, participating in a Hash House Harrier treasure hunt, or serving as a casual volunteer by, say, collecting bottles for the Scouts or making tea and coffee after a religious service. And Julia Harrison's (2001) study of upper-middle-class Canadian mass tourists revealed a certain level of complexity in their sensual experience of the touristic sites they visited. For people craving the simple things in life, this is the kind of leisure to head for. The other two domains abound with equivalent simple core activities, as in the work of a parking lot attendant (receiving cash/making change) or the efforts of a householder whose nonwork obligation of the day is raking leaves.

So, if complexity is what people want, they must look elsewhere. Leisure projects are necessarily more complex than casual leisure activities. The types of projects listed later in this chapter provide, I believe, ample proof of that. Nonetheless, they are not nearly as complex as the core activities around which serious leisure revolves. The accumulated knowledge, skill, training, and experience of, for instance, the amateur trumpet player, hobbyist stamp collector, and volunteer emergency medical worker are vast, and defy full description of how they are applied during conduct of the core activity. Of course, neophytes in the serious leisure activities lack these acquisitions, though it is unquestionably their intention to acquire them to a level where they will feel fulfilled. As with simple core activities complex equivalents also exist in the other two domains. Examples in work include the two earlier examples of the surgeon and jazz clarinetist. In the nonwork domain two examples considered later in this chapter are more or less complex: driving in city traffic and (for some people) preparing their annual income tax return.

The serious leisure perspective

The serious leisure perspective may be described, in simplest terms, as the theoretic framework that synthesizes three main forms of leisure showing, at once, their distinctive features, similarities, and interrelationships.[3] Additionally the Perspective (wherever Perspective appears as shorthand for serious leisure perspective, to avoid confusion, the first letter will be capitalized) considers how the three forms – serious leisure, casual leisure, and project-based leisure – are shaped by various psychological, social, cultural, and historical conditions. Each form serves as a conceptual umbrella for a range of types of related activities. That the Perspective takes its name from the first of these should, in no way, suggest that I regard it, in some abstract sense, as the most important or superior of the three. Rather the Perspective is so titled, simply because it got its start in the study of serious leisure; such leisure is, strictly from the standpoint of intellectual invention, the godfather of the other two. Furthermore, serious leisure has become the benchmark from which analyses of casual and project-based leisure have often been undertaken. So naming the Perspective after the first facilitates intellectual recognition; it keeps the idea in familiar territory for all concerned.

My research findings and theoretic musings over the past 35 years have nevertheless evolved and coalesced into a typological map of the world of leisure (for a brief history of the Perspective, see the history page at www.soci.ucalgary.ca/seriousleisure, or for a longer version, see Stebbins, 2007a, Chapter 6). That is, so far as known at present, all leisure (at least in Western society) can be classified according to one of the three forms and their several types and subtypes. More precisely the serious leisure perspective offers a classification and explanation of all leisure activities and experiences, as these two are framed in the social

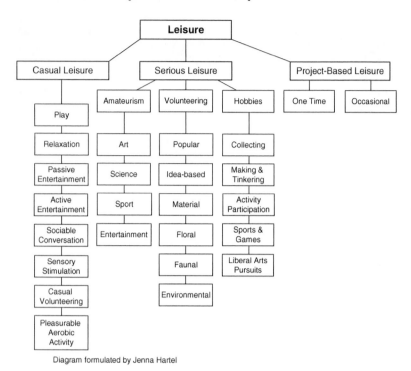

Diagram formulated by Jenna Hartel

Figure 1.1 The Serious Leisure Perspective
Source: http://www.soci.ucalgary.ca/seriousleisure

psychological, social, cultural, and historical conditions in which each activity and accompanying experience take place. Figure 1.1 portrays the typological structure of the Perspective.

Serious leisure

Serious leisure is the systematic pursuit of an amateur, hobbyist, or volunteer activity sufficiently substantial, interesting, and fulfilling for the participant to find a (leisure) career there acquiring and expressing a combination of its special skills, knowledge, and experience. I coined the term (Stebbins, 1982) to express the way the people he interviewed and observed viewed the importance of these three kinds of activity in their everyday lives. The adjective 'serious' (a word my research respondents often used) embodies such qualities as earnestness, sincerity, importance, and carefulness, rather than gravity, solemnity, joylessness, distress, and anxiety. Although the second set of terms occasionally describes serious leisure events, they are uncharacteristic of them and fail to nullify, or, in many cases, even dilute, the overall fulfillment gained by the participants. The idea of 'career' in this definition follows sociological tradition, where careers are seen as available in all substantial, complex roles, including those in leisure. Finally, as we shall see shortly, serious leisure is distinct from casual leisure and project-based leisure.

Amateurs are found in art, science, sport, and entertainment, where they are invariably linked in a variety of ways with professional counterparts. The two can be distinguished descriptively in that the activity in question constitutes a livelihood for professionals but not amateurs. Furthermore, most professionals work full-time at the activity whereas all amateurs pursue it part-time. The part-time professionals in art and entertainment complicate this picture; although they work part-time, their work is judged by other professionals and by the amateurs as of professional quality. Amateurs and professionals are locked in and therefore defined by a system of relations linking them and their publics – the 'professional-amateur-public system,' or P-A-P system (discussed in more detail in Stebbins, 1979, 1992, Chapter 3; 2007a, pp. 6–8).

Hobbyists lack this professional alter ego, suggesting that, historically, all amateurs were hobbyists before their fields professionalized. Both types are drawn to their leisure pursuits significantly more by self-interest than by altruism, whereas volunteers engage in activities requiring a more or less equal blend of these two motives. Hobbyists may be classified into five types: collectors, makers and tinkerers, noncompetitive activity participants (e.g., fishing, hiking, orienteering), hobbyist sports and games (e.g., ultimate Frisbee, croquet, gin rummy), and the liberal arts hobbies (primarily reading in an area of history, science, philosophy, literature, etc.; see Stebbins, 1994).

Smith, Stebbins and Dover (2006, pp. 239–240) define *volunteer* – whether economic or volitional – as someone who performs, even for a short period of time, volunteer work in either an informal or a formal setting. It is through volunteer work that this person provides a service or benefit to one or more individuals (they must be outside that person's family), usually receiving no pay, even though people serving in volunteer programs are sometimes compensated for out-of-pocket expenses. Moreover, in the field of nonprofit studies, since no volunteer work is involved, giving (of, say, blood, money, clothing), as an altruistic act, is not considered volunteering. Meanwhile, in the typical case, volunteers who are altruistically providing a service or benefit to others are themselves also benefiting from various rewards experienced during this process (e.g., pleasant social interaction, self-enriching experiences, sense of contributing to nonprofit group success). In other words, volunteering is motivated by two basic attitudes: altruism *and* self-interest.

The conception of volunteering that squares best with a positive sociology revolves, in significant part, around a central subjective motivational question: it must be determined whether volunteers feel they are engaging in an enjoyable (casual leisure), fulfilling (serious leisure), or either enjoyable or fulfilling (project-based leisure) core activity that they have had the option to accept or reject on their own terms. A key element in the leisure conception of volunteering is the felt absence of coercion, moral or otherwise, to participate in the volunteer activity (Stebbins, 1996a), an element that, in 'marginal volunteering' (Stebbins, 2001b), may be experienced in degrees, as more or less coercive. The reigning conception of volunteering in nonprofit sector research is not that of volunteering as leisure, but rather volunteering as unpaid work. The first – an *economic* conception – defines volunteering as the absence of payment as livelihood, whether in money or in kind. This definition, for the most part, leaves unanswered the messy question of motivation so crucial to the second, positive sociological, definition, which is a *volitional* conception.

Volitionally speaking, volunteer activities are motivated, in part, by one of six types of interest: interest in activities involving (1) people, (2) ideas, (3) things, (4) flora, (5) fauna, or (6) the natural environment (Stebbins, 2007b). Each type, or combination of types, offers its volunteers an opportunity to pursue, through an altruistic activity, a particular kind of interest. Thus, volunteers interested in working with certain ideas are attracted to idea-based volunteering, while those interested in certain kinds of animals are attracted to faunal volunteering. Interest forms the first dimension of a typology of volunteers and volunteering.

But, since volunteers and volunteering cannot be explained by interest alone, a second dimension is needed. This is supplied by the serious leisure perspective and its three forms. This perspective, as already noted, sets out the motivational and contextual (sociocultural, historical) foundation of the three. The intersections of these two dimensions produce 18 types of volunteers and volunteering, exemplified in idea-based serious leisure volunteers, material casual leisure volunteering (working with things), and environmental project-based volunteering (see Table 1.1)

Table 1.1   A Leisure-Based Theoretic Typology of Volunteers and Volunteering

Leisure Interest	Type of Volunteer		
	Serious Leisure (SL)	Casual Leisure (CL)	Project-Based Leisure (PBL)
Popular	SL Popular	CL Popular	PBL Popular
Idea-Based	SL Idea-Based	CL Idea-Based	PBL Idea-Based
Material	SL Material	CL Material	PBL Material
Floral	SL Floral	CL Floral	PBL Floral
Faunal	SL Faunal	CL Faunal	PBL Faunal
Environmental	SL Environmental	CL Environmental	PBL Environmental

Six qualities

Serious leisure is further defined by six distinctive qualities, qualities uniformly found among its amateurs, hobbyists, and volunteers. One is the occasional need to *persevere*. Participants who want to continue experiencing the same level of fulfillment in the activity have to meet certain challenges from time to time. Thus, musicians must practise assiduously to master difficult musical passages, baseball players must throw repeatedly to perfect favorite pitches, and volunteers must search their imaginations for new approaches with which to help children with reading problems. It happens in all three types of serious leisure that deepest fulfillment sometimes comes at the end of the activity rather than during it, from sticking with it through thick and thin, from conquering adversity.

Another quality distinguishing all three types of serious leisure is the opportunity to follow a (leisure) *career* in the endeavor, a career shaped by its own special contingencies, turning points, and stages of achievement and involvement. A career that, in some fields notably certain arts and sports, may nevertheless include decline. Moreover, most, if not all, careers here owe their existence to a third quality: serious leisure participants make significant personal *effort* using their specially acquired knowledge, training, or skill and, indeed at times, all three. Careers for serious leisure participants unfold along lines of their efforts to achieve, for instance, a high level of showmanship, athletic prowess, or scientific knowledge or to accumulate formative experiences in a volunteer role.

Serious leisure is further distinguished by several *durable benefits*, or tangible, salutary outcomes such activity for its participants. They include self-actualization, self-enrichment, self-expression, regeneration or renewal of self, feelings of accomplishment, enhancement of self-image, social interaction and sense of belonging, and lasting physical products of the activity (e.g., a painting, scientific paper, piece of furniture). A further benefit – self-gratification, or pure fun, which is by far the most evanescent benefit in this list – is also enjoyed by casual leisure participants. The possibility of realizing such benefits constitutes a powerful goal in serious leisure.

Fifth, serious leisure is distinguished by a unique *ethos* that emerges in parallel with each expression of it. An ethos is the spirit of the community of serious leisure participants, as manifested in shared attitudes, practices, values, beliefs, goals, and so on. The social world of the participants is the organizational milieu in which the associated ethos – at bottom a cultural formation – is expressed (as attitudes, beliefs, values) or realized (as practices, goals). According to David Unruh (1979, 1980) every social world has its characteristic groups, events, routines, practices, and organizations. It is held together, to an important degree, by semi-formal, or mediated, communication. In other words, in the typical case, social worlds are neither heavily bureaucratized nor substantially organized through intense face-to-face interaction. Rather, communication is commonly mediated by newsletters, posted notices, telephone messages, mass mailings, radio and television announcements, and similar means.

The social world is a diffuse, amorphous entity to be sure, but nevertheless one of great importance in the impersonal, segmented life of the modern urban community. Its importance is further amplified by a parallel element of the special ethos, which is missing from Unruh's conception, namely that such worlds are also constituted of a rich subculture. One function of this subculture is to interrelate the many components of this diffuse and amorphous entity. In other words, there is associated with each social world a set of special norms, values, beliefs, styles, moral principles, performance standards, and similar shared representations.

Every social world contains four types of members: strangers, tourists, regulars, and insiders (Unruh, 1979, 1980). The strangers are intermediaries who normally participate little in the leisure activity itself, but who nonetheless do something important to make it possible, for example, by managing municipal parks (in amateur baseball), minting coins (in hobbyist coin collecting), and organizing the work of teachers' aids (in career volunteering). Tourists are temporary participants in a social world; they have come on the scene momentarily for entertainment, diversion, or profit. Most amateur and hobbyist activities have publics of some kind, which are, at bottom, constituted of tourists. The clients of many volunteers can be similarly classified. The regulars routinely participate in the social world; in serious leisure, they are the amateurs, hobbyists, and volunteers themselves. The insiders are those among them who show exceptional devotion to the social world they share, to maintaining it, to advancing it. In the serious leisure perspective such people are analyzed according to an involvement scale as either 'core devotees' or 'moderate devotees' and contrasted with 'participants,' or regulars (Siegenthaler and O'Dell, 2003; Stebbins, 2007a, pp. 20–21).

The sixth quality – participants in serious leisure tend to identify strongly with their chosen pursuits – springs from the presence of the other five distinctive qualities. In contrast, most casual leisure, although not usually humiliating or despicable, is nonetheless too fleeting, mundane, and commonplace to become the basis for a distinctive *identity* for most people.

Motivation

Furthermore, certain rewards and costs come with pursuing a hobbyist, amateur, or volunteer activity. Both implicitly and explicitly much of serious leisure theory rests on the following assumption: to understand the meaning of such leisure for those who pursue it is in significant part to understand their motivation for the pursuit. Moreover, one fruitful approach to understanding the motives that lead to serious leisure participation is to study them through the eyes of the participants who, past studies reveal (Stebbins, 1992, Chapter 6; 1996b; 1998; Arai and Pedlar, 1997), see it as a mix of offsetting costs and rewards experienced in the central activity. The rewards of this activity tend to outweigh the costs, however, the result being that the participants usually find a high level of personal fulfillment in them.

The rewards of a serious leisure pursuit are the more or less routine values that attract and hold its enthusiasts. Every serious leisure career both frames and is framed by the continuous search for these rewards, a search that takes months, and in some fields years, before the participant consistently finds deep satisfaction in his or her amateur, hobbyist, or volunteer role. Ten rewards have so far emerged in the course of my various exploratory studies of amateurs, hobbyists, and career volunteers. As the following list shows, the rewards are predominantly personal.

Personal rewards

1. Personal enrichment (cherished experiences).
2. Self-actualization (developing skills, abilities, knowledge).
3. Self-expression (expressing skills, abilities, knowledge already developed).
4. Self-image (known to others as a particular kind of serious leisure participant).
5. Self-gratification (combination of superficial enjoyment and deep fulfillment).
6. Re-creation (regeneration) of oneself through serious leisure after a day's work.
7. Financial return (from a serious leisure activity).

Social rewards

8. Social attraction (associating with other serious leisure participants, with clients as a volunteer, participating in the social world of the activity).
9. Group accomplishment (group effort in accomplishing a serious leisure project; senses of helping, being needed, being altruistic).
10. Contribution to the maintenance and development of the group (including senses of helping, being needed, being altruistic in making the contribution).

Further, every serious leisure activity contains its own costs – a distinctive combination of tensions, dislikes, and disappointments – which each participant confronts in his or her special way. Tensions and dislikes develop within the activity or through its imperfect mesh with work, family, and other leisure interests. Put more precisely, the goal of gaining fulfillment in serious leisure is the drive to experience the rewards of a given leisure activity, such that its costs are seen by the participant as more or less insignificant by comparison. This is at once the meaning of the activity for the participant and that person's motivation for engaging in it. It is this motivational sense of the concept of reward that distinguishes it from the idea of durable benefit set out earlier, an idea that emphasizes outcomes rather than antecedent conditions.

Nonetheless, the two ideas constitute two sides of the same social psychological coin. Moreover, this brief discussion shows that some positive psychological states may be founded, to some extent, on particular negative, often noteworthy, conditions (e.g., tennis elbow, frostbite [cross-country skiing], stage fright, frustration [in acquiring a collectable, learning a part]). Such conditions can make the senses of achievement and self-fulfillment even more pronounced as the enthusiast manages to conquer adversity.

Thrills and psychological flow

Thrills are part of this reward system. *Thrills*, or high points, are the sharply exciting events and occasions that stand out in the minds of those who pursue a kind of serious leisure or devotee work. In general, they tend to be associated with the rewards of self-enrichment and, to a lesser extent, those of self-actualization and self-expression. That is, thrills in serious leisure and devotee work may be seen as situated manifestations of certain more abstract rewards; they are what participants in some fields seek as concrete expressions of the rewards they find there. They are important, in substantial part, because they motivate the participant to stick with the pursuit in the hope of finding similar experiences again and again and because they demonstrate that diligence and commitment may pay off. Because thrills, as defined here, are based on a certain level of mastery of a core activity, they know no equivalent in casual leisure. The thrill of a roller coaster ride is qualitatively different from a successful descent down a roaring rapids in a kayak where the boater has the experience, knowledge, and skill to accomplish this.

Over the years I have identified a number of thrills that come with the serious leisure activities I studied. These thrills are exceptional instances of the *flow* experience. Thus, although the idea of flow originated with the work of Mihalyi Csikszentmihalyi (1990), and has therefore an intellectual history quite separate from that of serious leisure, it does nevertheless happen, depending on the activity, that it is a key motivational force there. For example, I found flow was highly prized in the hobbies of kayaking, mountain/ice climbing, and snowboarding (Stebbins, 2005b). What then is flow?

The intensity with which some participants approach their leisure suggests that, there, they may at times be in psychological flow. Flow, a form of optimal experience, is possibly the most widely discussed and studied generic intrinsic reward in the psychology of work and leisure. Although many types of work and leisure generate little or no flow for their participants, those that do are found primarily the 'devotee occupations' (discussed in Chapter 5) and serious leisure. Still, it appears that each work and leisure activity capable of producing flow does so in terms unique to it. And it follows that each of these activities, especially their core activities, must be carefully studied to discover the properties contributing to the distinctive flow experience it offers.

In his theory of optimal experience, Csikszentmihalyi (1990, pp. 3–5, 54) describes and explains the psychological foundation of the many flow activities in work and leisure, as exemplified in chess, dancing, surgery, and rock climbing. Flow is 'autotelic' experience, or the sensation that comes with the actual enacting of intrinsically rewarding activity. Over the years Csikszentmihalyi (1990, pp. 49–67) has identified and explored eight components of this experience. It is easy to see how this quality of complex core activity, when present, is sufficiently rewarding and, it follows, highly valued to endow it with many of the qualities of serious leisure, thereby rendering the two, at the motivational level, inseparable in several ways. And this holds even though most people tend to think of work and leisure as vastly different. The eight components are

1. sense of competence in executing the activity;
2. requirement of concentration;
3. clarity of goals of the activity;
4. immediate feedback from the activity;
5. sense of deep, focused involvement in the activity;
6. sense of control in completing the activity;
7. loss of self-consciousness during the activity;
8. sense of time is truncated during the activity.

These components are self-evident, except possibly for the first and the sixth. With reference to the first, flow fails to develop when the activity is either too easy or too difficult; to experience flow the participant must feel capable of performing a moderately challenging activity. The sixth component refers to the perceived degree of control the participant has over execution of the activity. This is not a matter of personal competence; rather it is one of degree of maneuverability in the fact of uncontrollable external forces, a condition well illustrated in situations faced by the mountain hobbyists mentioned above, as when the water level suddenly rises on the river or an unpredicted snowstorm results in a white-out on a mountain snowboard slope. Viewed from the serious leisure perspective psychological flow tends to be associated with the rewards of self-enrichment and, to a lesser extent, those of self-actualization and self-expression.

Casual leisure

Casual leisure is immediately intrinsically rewarding, relatively short-lived pleasurable activity requiring little or no special training to enjoy it. It is fundamentally hedonic, pursued for its significant level of pure enjoyment, or pleasure. Much of the debate over the years on mass leisure revolves around this form. The termed was coined by the author in the 1982 conceptual statement about serious leisure, which at the time, depicted its casual counterpart as all activity not classifiable as serious leisure (project-based leisure has since been added as a third form, see next section). As a scientific concept casual leisure languished in this residual status, until Stebbins (1997a, 2001c), belatedly recognizing its centrality and importance in leisure studies, sought to elaborate the idea as a sensitizing concept for exploratory research, as he had earlier for serious leisure (see also Rojek, 1997). It is considerably less substantial and offers no career of the sort found in serious leisure.

Its types – there are eight (see Figure 1.1) – include *play* (including dabbling), *relaxation* (e.g., sitting, napping, strolling), *passive entertainment* (e.g., TV, books, recorded music), *active entertainment* (e.g., games of chance, party games), *sociable conversation, sensory stimulation* (e.g., sex, eating, drinking), and *casual volunteering* (as opposed to serious leisure, or career, volunteering). Note that dabbling (as play) may occur in the same genre of activity pursued by amateurs, hobbyists, and career volunteers. The preceding section was designed, in part, to conceptually separate dabblers from this trio of leisure participants, thereby enabling the reader to interpret with sophistication references to, for example, 'amateurish' activity (e.g., *The Cult of the Amateur* by Andrew Keen, 2007).

The last and newest type of casual leisure – *pleasurable aerobic activity* – refers to physical activities that require effort sufficient to cause marked increase in respiration and heart rate. Here I am referring to 'aerobic activity' in the broad sense, to all activity that calls for such effort, which to be sure, includes the routines pursued collectively in (narrowly conceived of) aerobics classes and those pursued individually by way of televised or video-taped programs of aerobics (Stebbins, 2004b). Yet, as with its passive and active cousins in entertainment, pleasurable aerobic activity is basically casual leisure. That is, to do such activity requires little more than minimal skill, knowledge, or experience. Examples include the game of the Hash House Harriers (a type of treasure hunt in the outdoors), kickball (described in *The Economist*, 2005, as a cross between soccer and baseball), and such children's games as hide-and-seek.

It is likely that people pursue the different types of casual leisure in combinations of two and three at least as often as they pursue them separately. For instance, every type can be relaxing, producing in this fashion play-relaxation, passive entertainment-relaxation, and so on. Various combinations of play and sensory stimulation are also possible, as in experimenting, in deviant or nondeviant ways, with drug use, sexual activity, and thrill seeking through movement. Additionally, sociable conversation accompanies some sessions of sensory stimulation (e.g., recreational drug use, curiosity seeking, displays of beauty) as well as some sessions of relaxation and active and passive entertainment, although such conversation normally tends to be rather truncated in the latter two.

Notwithstanding its hedonic nature casual leisure is by no means wholly inconsequential, for some clear costs and benefits accrue from pursuing it. Moreover, in contrast to the evanescent hedonic property of casual leisure itself, these costs and benefits are enduring. The benefits include serendipitous creativity and discovery in play, regeneration from early intense activity, and development and maintenance of interpersonal relationships (Stebbins, 2001c; other benefits are discussed in Stebbins, 2007a, pp. 41–43). Some of its costs root in excessive casual leisure or lack of variety as manifested in boredom or lack of time for leisure activities that contribute to self through acquisition of skills, knowledge, and experience (i.e., serious leisure). Moreover, casual leisure alone is unlikely to produce a distinctive leisure identity.

Project-based leisure

Project-based leisure (Stebbins, 2005c) is the third form of leisure activity and the most recent one added to the Perspective. It is a short-term, reasonably complicated, one-off or occasional, though infrequent, creative undertaking carried out in free time, or time free of disagreeable obligation. Such leisure requires considerable planning, effort, and sometimes skill or knowledge, but is for all that neither serious leisure nor intended to develop into such. Examples include surprise birthday parties, elaborate preparations for a major holiday, and volunteering for sports events. Though only a rudimentary social world springs up around the project, it does, in its own particular way, bring together friends, neighbors, or relatives (e.g., through a genealogical project or Christmas celebrations), or draw the individual participant into an organizational milieu (e.g., through volunteering for a sports event or major convention).

Types of project-based leisure

It was noted in the definition just presented that project-based leisure is not all the same. Whereas systematic exploration may reveal others, two types are evident at this time: one-off projects and occasional projects. These are presented next using the classificatory framework for amateur, hobbyist, and volunteer activities developed earlier in this chapter.

One-off projects

In all these projects people generally use the talents and knowledge they have at hand, even though for some projects they may seek certain instructions beforehand, including reading a book or taking a short course. And some projects resembling hobbyist activity participation may require a modicum of preliminary conditioning. Always, the goal is to undertake successfully the one-off project and nothing more, and sometimes a small amount of background preparation is necessary for this. It is possible that a survey would show that most project-based leisure is hobbyist in character and the next most common, a kind of volunteering. First, the following hobbyist-like projects have so far been identified:

- Making and tinkering:
    - Interlacing, interlocking, and knot-making from kits.
    - Other kit assembly projects (e.g., stereo tuner, craft store projects).
    - Do-it-yourself projects done primarily for fulfillment, some of which may even be undertaken with minimal skill and knowledge (e.g., build a rock wall or a fence, finish a room in the basement, plant a special garden). This could turn into an irregular series of such projects, spread over many years, possibly even transforming the participant into a hobbyist.

- Liberal arts:
    - Genealogy (not as ongoing hobby).
    - Tourism: special trip, not as part of an extensive personal tour program, to visit different parts of a region, a continent, or much of the world.

- Activity participation: long back-packing trip, canoe trip; one-off mountain ascent (e.g., Fuji, Rainier, Kilimanjaro).

One-off volunteering projects are also common, though possibly somewhat less so than hobbyist-like projects. And less common than either are the amateur-like projects, which seem to concentrate in the sphere of theater.

- Volunteering:

    o Volunteer at a convention or conference, whether local, national, or international in scope.

    o Volunteer at a sporting competition, whether local, national, or international in scope.

    o Volunteer at an arts festival or special exhibition mounted in a museum.

    o Volunteer to help restore human life or wildlife after a natural or human-made disaster caused by, for instance, a hurricane, earthquake, oil spill, or industrial accident.

- Entertainment Theater: produce a skit (a form of sketch) or one-off community pageant; create a puppet show; prepare a home film or a set of videos, slides, or photos; prepare a public talk.

Occasional projects

The occasional projects seem more likely to originate in or be motivated by agreeable obligation than their one-off cousins. Examples of occasional projects include the sum of the culinary, decorative, or other creative activities undertaken, for example, at home or at work for a religious occasion or for someone's birthday. Likewise, national holidays and similar celebrations sometimes inspire individuals to mount occasional projects consisting of an ensemble of inventive elements.

Unlike one-off projects occasional projects have the potential to become routinized, which happens when new creative possibilities no longer come to mind as the participant arrives at a fulfilling formula wanting no further modification. North Americans who decorate their homes the same way each Christmas season exemplify this situation. Indeed, it can happen that, over the years, such projects may lose their appeal, but not their necessity, thereby becoming disagreeable obligations, which their authors no longer define as leisure.

And, lest it be overlooked, note that one-off projects also hold the possibility of becoming unpleasant. Thus, the hobbyist genealogist gets overwhelmed with the details of family history and the challenge of verifying dates. The thought of putting in time and effort doing something once considered leisure but which she now dislikes makes no sense. Likewise, volunteering for a project may turn sour, creating in the volunteer a sense of being faced with a disagreeable obligation, which however, must still be honored. This is leisure no more.

Deviant leisure

The study of deviant leisure is at least a decade old (see Stebbins, 1996d, 1997a; Rojek, 1997, pp. 392–393; 2000, Chapter 4; Cantwell, 2003; special issue of *Leisure/Loisir*, vol. 30, no. 1, 2006), and readers interested in it are encouraged to turn to these sources. What is important to note with respect to the serious leisure perspective is that deviant leisure may take either the casual or the serious form (we have so far been unable to identify any project-based deviant leisure). Casual leisure is probably the more common and widespread of the two.

Casual or serious, deviant leisure mostly fits the description of 'tolerable deviance' (exceptions are discussed below). Although its contravention of certain moral norms of a society is held by most of its members to be mildly threatening in most social situations, this form of deviance nevertheless fails to generate any significant or effective communal attempts to control it (Stebbins, 1996c, pp. 3–4). Tolerable deviance undertaken for pleasure – as casual leisure – encompasses a range of deviant sexual activities including cross-dressing, homosexuality, watching sex (e.g., striptease, pornographic films), and swinging and group sex. Heavy drinking and gambling, but not their more seriously regarded cousins alcoholism and compulsive gambling, are also tolerably deviant forms of casual leisure, as are the use of cannabis and the illicit, pleasurable, use of certain prescription drugs. Social nudism has also been analyzed within the tolerable deviance perspective (all these forms are examined in greater detail with accent on their leisure qualities in Stebbins, 1996c, Chapters 3–7, 9).

In the final analysis, deviant casual leisure roots in sensory stimulation and, in particular, the creature pleasures it produces. The majority of people in society tolerate most of these pleasures even if they would never think, or at least not dare, to enjoy themselves in these ways. In addition, they actively scorn a somewhat smaller number of intolerable forms of deviant casual leisure, demanding decisive police control of, for example, incest, vandalism, sexual assault, and what Jack Katz (1988, Chapter 2) calls the 'sneaky thrills' (certain incidents of theft, burglary, shoplifting, and joyriding).[4] Sneaky thrills, however, are motivated not by the desire for creature pleasure, but rather by the desire for a special kind of excitement, namely, going against the grain of established social life.

Beyond the broad domains of tolerable and intolerable deviant casual leisure lies that of deviant serious leisure, composed primarily of aberrant religion, politics, and science. Deviant religion is manifested in the sects and cults of the typical modern society, while deviant politics is constituted of the radical fringes of its ideological left and right. Deviant science centers on the occult which, according to Marcello Truzzi (1972), consists of five types: divination, witchcraft–Satanism, extrasensory perception, Eastern religious thought, and various residual occult phenomena revolving around unidentified flying objects (UFOs), water witching, lake monsters, and the like (for further details, see Stebbins, 1996c, chapter 10). Thus deviant serious leisure, in the main, is pursued as a liberal arts hobby or as activity participation, or in fields like witchcraft and divination, as both.

In whichever form of deviant serious leisure a person participates, he or she will find it necessary to make a significant effort to acquire its special belief system as well as to defend it against attack from mainstream science, religion, or politics. Moreover, here, the person will discover two additional rewards of considerable import: a special personal identity grounded, in part, in the unique genre of self-enrichment that invariably comes with inhabiting any marginal social world.

Clearly there are various kinds of deviant leisure (most of it serious leisure) that fall beyond the orbit of the commercial consumption of goods and services, among them, joyriding, swinging at private parties, and much of deviant science. Yet a good amount of casual deviant leisure *is* consumptive, as in buying pornographic videos, illegal recreational drugs and alcohol (for the alcoholic), and the services of a prostitute. Indeed deviance offers a fine laboratory for exploring both the separateness and the overlap of consumption and leisure.

Conclusions

The two main sections of this chapter – the nature of consumption and the nature of leisure – contrast with one another in that, generally speaking, the first is macro-sociological whereas the second is micro-sociological. In the first we considered leisure society, some of its trends, and some of the debate about what they mean. In the second we stayed largely on the experiential level (with the partial exception of social world), discussing such processes as activity, motivation, career, flow, and deviance. Still I did state near the beginning of the section on the serious leisure perspective that its three forms – serious leisure, casual leisure, and project-based leisure – are shaped by various psychological, social, cultural, and historical conditions. The macro-sociological section preceding it is precisely about such conditions.

The next two chapters present considerably more detail about these background conditions, served up headings of conspicuous consumption and consumptive leisure in context. Then, in Chapters 4 and 5, which set out Phases One and Two of consumption as shopping and then as consuming the purchase, we shall see how the micro-sociological part of the consumptive life as it relates to leisure fits within this macro-sociological shell. Do not expect the fit to be perfect in all respects, for it will not. These two grand theoretic-analytic traditions have a sad history of paying rather little attention to each other. Chapter 6 returns to the contextual motif with an examination of the organizational basis of leisure and consumption. It ends on a micro-macro note, however, with a look at two critical issues raised by consumptive leisure: the drive for voluntary simplicity and leisure's impact on the environment.

2
Conspicuous Consumption

This chapter explores the ideas of Thorstein Veblen, Marcel Mauss, and others as originally written and then applies these ideas to consumer life in the 21st century. Various principles of the serious leisure perspective are used to link conspicuous consumption in free time. Conspicuous consumption elevates significantly the importance for the consumer of the commercial side of this person's leisure. Purchasing expensive, dazzling goods and services earns the buyer a cachet in the eyes of the other people in that person's circle. As Veblen (1899, p. 64) put it in his day: 'conspicuous consumption of valuable goods is a means of reputability to the gentleman of leisure.'

We begin with a review of past and present thought on conspicuous consumption, gifts, and reciprocity. Then we examine consumption as leisure and identity, and conversely, the implications of leisure for such consumption.

Veblen's theory

Thorstein Veblen, an American economist who wrote mostly between 1899 and 1923, is generally credited with having pioneered the idea of conspicuous consumption. In his celebrated work, *The Theory of the Leisure Class* (Veblen, 1899), he argued that leisure can be used to demonstrate status and power in modern industrial society, such that this practice resulted in a distinctive leisure class. Nevertheless use of wealth in this way is of relatively recent origin. Historically societies were unable to produce a level of material goods beyond that needed for subsistence. But eventually parts of some societies – the main examples being the industrial and postindustrial societies of today – came to enjoy a surplus of these goods, which raised questions about how the surplus is controlled, distributed, and used. Control, distribution, and use refer, in effect, to sets of options, one of which is conspicuous consumption. Here use is manifested in the purchase and hence ownership of distinctive goods and services available only to people who have the money (control) to buy them (distribution).

A main process underlying conspicuous consumption is in Veblen's words 'invidious comparison.' It is based on ownership of distinctive consumer goods and services, or goods and services consensually recognized as indicators of their owner's accumulated wealth. This consensus was, in Veblen's day, shared by the wealthy elite, and it centered on having the taste to make the 'right' acquisitions. In other words, it is sometimes possible to pay a great deal for an object or a service, but still fail to buy one regarded by this elite as demonstrating financial and cultural sophistication. Consider a modern example. Two wealthy people spend 10 million dollars to have their mansions built, but one of them (the less sophisticated) has tastes for interior design that are judged by leisure-class standards as 'plebian,' tastes that compare poorly with those of the other (more sophisticated) buyer.

The leisure class for whom conspicuous consumption was a mark of membership also embraced the value of being exempt from all necessary remunerated employment. Not having to work for a living is honorable and meritorious – the very essence of a decent person. This orientation toward work has ancient roots, most notably in the writings of Plato and Aristotle. In Veblen's theory leisure connotes the nonproductive use of free time. Nevertheless leisure for the leisure class was often, in accord with the Perspective presented earlier, of the serious variety, exemplified in learning ancient languages, studying occult science, breeding show horses, and going in for equestrian sports. Good manners and refined tastes were considered the *sine qua non* of the properly cultured and properly leisured gentlemen, which however, like all serious leisure, took time and perseverance to learn and perfect.

For members of the leisure class the principle of conspicuous consumption tended to penetrate all of life, even those parts hidden from public view. As Veblen (1899, p. 46) put it: 'leisure ... does not connote indolence or quiescence. What it connotes is non-productive consumption of time. Time is consumed non-productively.' He observed further that some of this nonproductive time was devoted to leisure pursued privately. Nonetheless the gentlemen of leisure must be able to 'give a convincing account' of how he has used this time, which might be done publicly through artistic expression (e.g., a painting at an exhibition), expression of knowledge (e.g., a published book), presentation of pure-bred animals (e.g., an entry in dressage competition), and similar strategies.

Purchases, even in the private sphere, had to be expensive and in good taste, including underwear, everyday cutlery, and household appliances. To be beautiful, honorable, and meritorious in the eyes of peers in this class, the good or service had also to be expensive. It was, therefore, something unavailable to people in the other classes. Not so hidden from public scrutiny was the pursuit of higher education by the wealthy for themselves and their children. Here, too, diplomas should be obtained from the most elite and expensive private schools, colleges, and universities.

Conspicuous consumption is alive and well in the 21st century. The process is now much broader than in the late 19th century, however, since today people sometimes conspicuously purchase goods or services for reasons other than free-time enjoyment, satisfaction, or fulfillment. Thus, some independent entrepreneurs, who must pay their own expenses, have elaborately decorated offices, travel on first-class air tickets, and drive expensive cars expressly because they want their clients to regard their businesses as successful. Such success, it is hoped, will be seen by present clients as built from serving earlier clients well. Nor is conspicuous consumption any longer the exclusive preserve of a leisure class, an observation that will be discussed in more detail later in this chapter.

Yet today, as in Veblen's time, conspicuous leisure continues to demonstrate that certain people have not only the wealth for expensive purchases (class) but also the taste to make respectable ones (status). Examples abound. Automobiles constitute an especially visible stage on which to strut this sort of display. Thus, the basic Ferrari, ranging in price from C$237,000 to 400,000, clearly signals both its owner's financial status and that person's taste in fine automotives. A business vice-president faced a dilemma along these lines over cars:

> Dear Tom and Ray:
> I am an older woman who has just taken a vice-presidential position with a company that is virtually all male and, thus, I do not want to display my profound ignorance about cars. The company has said that it will provide me with a car to use, BUT it must be one that reflects the company's success and status. Translated, that means I have to choose something fancier that the 2000 Lexis I currently rive
> I am, frankly, looking for comfort. I like leather seats, and I want roadside assistance if possible. One of the fellows suggested a Bentley,

but I don't know if the make something smaller than the Queen Mary (I like my car's size). Another fellow who collects cars says that if I were a car, I would be a classic Crown Vic. But a new one isn't good enough for my company.

So, here is the question: What car can I buy? I live in sunny California. I am told that Cadillacs and Lincolns are out. Something called an S-Class is OK on the low end (but I don't know what that is). Does Bentley make something of moderate Size? I am an old lady, and don't want to climb in and out of a sports car. Find me a car, please.
Thanks.
    Andrea
    (Magliozzi and Magliozzi, 2008)

Patricia Kranz (2008) writes about 'one-upmanship' among contemporary billionaires, clearly a consumptive league of its own. One vehicle some have chosen for conspicuous display, in addition to owning a private jet, a couple of mansions, and a Rolls-Royce or two, is having a magnificent yacht. Kranz reports that the competition in yachts shapes up along lines of length, with, as of March 2008, the longest boat – it measures 531.5 feet – still being in the works. It will be 6.5 feet longer than the Dubai, owned by Sheik Mohammed bin Rashid al-Maktoum, and to this date the longest yacht.

Primitive conspicuous consumption

Conspicuous consumption was not 'invented' by Veblen's leisure class, in that the practice had been observed earlier by anthropologists studying certain primitive societies. They found one manifestation of it in the *potlatch*, a lavish, competitive, ceremonial feast held in some of the groups of Northwest Pacific coast Indians. During a *potlatch* someone, usually a chief, gives presents and gives away or destroys (usually by burning) various possessions. At its most extravagant the goal of this custom is to enhance the prestige of a chief and his village, tribe, or other social unit. Commoners also hold *potlatches*, but they usually invite only local guests, whereas chiefs also invite people from many other tribes.

The practice amounts to a kind of conspicuous consumption, since as Veblen observed, 'as wealth accumulates on [the gentleman of leisure's] hands, his own unaided effort will not avail to sufficiently put his opulence in evidence by this method. The aid of friends and competitors is therefore brought in by resorting to the giving of valuable presents and expensive feasts and entertainments' (Veblen, 1899, pp. 64–65). The gentleman has purchased one or more of these, which his guests enjoy during their conspicuous display. The difference, historically, was that the chiefs gave away objects made by craftspeople in their tribe, objects not purchased in the marketplace.

According to the Peabody Museum at Harvard University, the *potlatch* is still held today. Contemporary gifts may be sums of money, however, rather than an object and possessions are not destroyed as a vehicle for display (http://www.peabody.harvard.edu/potlatch/default.html). Meanwhile these Indians are no longer primitive; they are native peoples living in a modern Western society.

A main purpose of the *potlatch* has always been the redistribution of wealth accumulated by families, a key means of political, economic, and social exchange. One may be held to celebrate a birth, puberty, marriage, death, or other important passage in status. Sometimes a *potlatch* honors an important person dead for many years. Although it is the feast that has attracted greatest attention – it is the purest manifestation of conspicuous consumption in this event – *potlatches* may also feature music, dance, drama, and spiritual ceremonies.

The *potlatch* serves to establish hierarchical relations within and between Indian clans and villages, as measured and reinforced by the distribution or destruction of wealth, presentation of dance and musical performances, and other ceremonial events. Family status in these societies is achieved not by the one with the most resources, but by the one who distributes the most of those resources. Dorothy Johansen describes further the process:

> In the *potlatch*, the host in effect challenged a guest chieftain to exceed him in his 'power' to give away or to destroy goods. If the guest did not return 100 percent on the gifts received and destroy even more wealth in a bigger and better bonfire, he and his people lost face and so his 'power' was diminished.
>
> (Johansen, 1967, pp. 7–8)

Mauss (1990) found a similar practice known as the *kula*, which was carried out by the Trobriand Islanders and their neighbors in the Massim in Northwest New Guinea. The term *kula* means circle, referring in this case to the ties with partners spread over the many islands and regions of Melanesia where they engaged in an international system of exchange.

This system consists of a set of local circles each large enough to intersect at its periphery with one or more neighboring circles.

> Participation in the *kula* represents the high point in the lives of Trobriand men. It is what wins them friends and renown. It makes life worth living and vies everything meaning. If we needed any proof of the superiority, in human affairs, of strictly symbolic motivations over those that are purely material, the impressive lasting power of the *kula* would be among the most eloquent. The *kula* exchange seems to have existed for at least five centuries.
>
> (Godbout, 1998, p. 106)

Nor is the *kula* of strictly historical interest; the practice remains to this day central to the lives of Trobriand Islanders and their neighbors, even in the face of westernization.

In the *kula*, as in the *potlatch*, luxury goods form the basis of exchange, and though utilitarian goods may also be exchanged, they must be endowed with a sense of luxury. To do otherwise is to show disrespect for the receiver of the gifts. Unlike the *potlatch* the *kula* is also prized for its tendency to establish friendships among the participants. They are in the exchange system, but may well live beyond the local circle. A chief might make as many as 200 friends in the course of exchanges conducted over several years. In both the *potlatch* and the *kula* barter is tolerated, but it is peripheral, occurs between episodes of ceremonial exchange, and is never practised by the nobility (Godbout, 1998, p. 106).

The gift

It was stated in the preceding chapter that receiving gifts falls outside the purview of this work, because giving one roots in the intention of the giver rather than that of the receiver. Nevertheless a giver may present a gift as, in whole or in part, an intentionally overt expression of consumptive power. Furthermore givers often purchase a good or service expressly for giving it to someone else. For the purposes of this book I define *giving* as an act of voluntarily transferring one or more goods or services to another person or group without expectation of anything in return. What is transferred is a *gift*. In remaining consistent with our theme of conspicuous consumption, the giver in this conceptualization is an individual or, at largest, a small group, but certainly nothing as big and impersonal as a large organization. So this is no place to discuss corporate giving.

These two definitions, viewed from the angle of conspicuous consumption, may be illustrated in many ways. Mary throws a lavish party to her friends, one goal of which is to prove to them she has the wherewithal to do this. John gives a colleague at work two free tickets in elite seating at a professional football match, his intention being to subtly underscore his financial capacity to buy high-priced season tickets to the local team's games. Mr and Mrs Murray give a million pounds as a charitable gift to a renowned private school, the money being earmarked for construction of a small concert hall, which will house a high-quality pipe organ. The hall and the organ will bear their name, and will trumpet to all who attend concerts there the fact of the Murrays' abundant wealth. Jack Brown buys a kidney and the set of medical and related services needed to transplant it in a young child. This charitable act draws extensive laudatory publicity in the local media, and Brown's star as a wealthy, but caring, member of the community rises immediately.

A notable point about these examples is that conspicuous display may be but one of several motives for giving something, a mental state perhaps most obvious in the example of Jack Brown. Veblen would argue that the display motive is nonetheless one of the most important, if not the most important, of all the motives operating in conspicuous gift giving. Marcel Mauss (1990, p. 47), whose anthropological work on the gift is renowned as a classic, recognized that, in the *potlatch*, the gifts given to guests were, in the language of this section, an expression of conspicuous display and not necessarily one of conspicuous consumption.

Furthermore we must not overlook the fact that purchasing a good or service to give to someone else is not always done with the aim of showing off the purchaser's wealth. Godbout (1998, p. 175) says that people sometimes give to others out of a sense of generosity or altruism, with demonstrations of superiority in personal riches and success being the farthest concern from their minds. Individual helping seems an especially good arena for this sort of largesse, as when a rich uncle takes pity on and then bails out a nephew in dire financial straits caused by excessive health care expenses. The uncle pays for, or consumes, the expenses in question as a gift to his nephew. In another example a daughter buys her elderly mother a new vacuum cleaner, one more efficient and easier to use and store than the present one. This gift may cost more than the daughter can comfortably afford, but compassion for her mother's situation overrides this financial worry. In other words gift giving is not always consumptive – gifts may be used items, donated services, or home-made objects the raw materials for which cost little or nothing (e.g., a home-made cake, knitted sweater, carved wooden toy) – and consumptive gift giving is not always motivated by conspicuous display of purchasing power, as seen in generously or altruistically giving a purchased good or service to someone.

Reciprocity

A gift, Mauss (1990, p. 45) wrote, tends to generate in the receiver a sense of reciprocity, a tendency to give something to the giver in return for what was received. George Homans (1974, p. 217) held that whatever is given in return should, according to the norm of reciprocity, be of equal value. The reciprocated item might be assessed for its equivalence along economic, symbolic, practical, or other lines. Thus, the Joneses, having been invited to a private dinner party, bring a gift of a bottle of wine. In effect this gift is in return for the pleasures of the evening, including especially the meal and its preparation. This gift, assuming the wine is of mutually agreed upon quality, has nevertheless symbolic equivalency. While its economic value is unlikely to equal the time, effort, and money the host put in to make the party a success. Nor does this bottle of wine have practical value (unless the host's supply inconveniently runs out and the gift is opened to make up for the shortfall).

Alvin Gouldner (1960) looked on reciprocity as a 'generalized moral norm,' hypothesizing that it constitutes a principal rule in the moral code of every society. He looked at reciprocity from several angles: institutional, interpersonal, group, and role, all of which are evident in the *potlatch*. By way of example consider the chief whose role is to host the *potlatch*, itself an institutionalized practice, who is invoking an interpersonal comparison with another chief and an intergroup comparison between their tribes. The gift of wine described in the preceding paragraph can be framed similarly: in refined circles in Western society, it is expected (an institutionalized norm) that guests coming to dinner will bring a token of their appreciation – a gift – a practice which does, however, trigger a tendency to evaluate both the evening's experience and the bottle of wine. This encourages a comparison between two groups (e.g., dyadic couples) or individuals.

As just observed for gifts, in general, a token of appreciation for a dinner party need not be purchased (e.g., a plate of home-baked cookies or bouquet of flowers from the giver's garden). Still, gifts are often precisely that – purchases – and, as such, may be vehicles for conspicuous display. So the Jones, hoping to show their sophistication as oenophiles as well as their economic standing, give a costly, prestigious, vintage French red to their hosts. And more generally now, what place does conspicuous consumption and display in gift giving and reciprocity have in the 21st century? How often do we, as givers, think about wedding, birthday, and Christmas gifts as signs of our taste and financial ranking?

The illustrations just presented suggest a couple of generalizations about conspicuous consumption and leisure in modern times. One of them is that routine consumption for the purposes of personal display as part of the lifestyle of a leisure class has not abated since Veblen wrote over 100 years ago. Another is that change has occurred during this period, for such consumption is now also evident outside that class, among the working wealthy and, on a reduced scale, among some more ordinary people. Today the latter have enough money to buy certain goods and services the cost and level of sophistication of which elevates their high status in the group. For instance, in some circles certain mobile phones, automobiles, items of apparel, tourist destinations, and brands of sporting equipment call attention to their owners as people of means and taste.

Leisure and conspicuous consumption today

This chapter has already served up a number of examples of present-day conspicuous consumption in the West, sufficient to belie any notion that the practice was limited to Veblen's time. The aim of the present section and the following one on identity is to underscore its contemporary existence and importance in relation to leisure. That leisure is an arena within which conspicuous display is sometimes observed emphasizes once again the value of considering consumption and leisure together. Of the many areas of modern life where leisure and conspicuous consumption meet, none illustrates this relationship better than the adoption of new technology.

Technology and consumption

Technology, more so today than ever, is a vast field strewn with new inventions of objects and processes. All areas considered – engineering, computing, aeronautics, automotive, medical, environmental, and others – the current rate of invention is staggeringly high. Still the sheer existence of something new does not inevitably lead to either quick or widespread adoption by targeted adopters. More precisely with many inventions there are early adopters, later adopters, and nonadopters, or people who reject the newcomer.

Early adopters commonly pay a higher price for the new item or process than later adopters, primarily because, by the time the second enter the market, the pool of possible buyers is typically much larger. This reflects an economy of scale at the production level, wherein certain earlier start-up costs of the product, including those of its development, are offset by later sales at reduced prices to a substantially greater number of buyers (for an example of this pricing cycle for video games, *The Economist*, 2008).

Thus, given the higher cost of the product at initial entry into the market and the prestige that often comes with owning or using the latest technology, the stage is set for a distinctive kind of modern-day conspicuous consumption. Early adopters may have to pay somewhat more for a recent invention, but this demonstrates financial capacity to do so and allows them to simultaneously bask in the warm adulation of other buyers who also fancy owning it. On a much more ordinary level than the one-upmanship of billionaires described earlier, such display is evident among early adopters of the latest technological advance in cameras, mobile phones, personal digital assistants, television screens, laptop computers, computer software, and on and on, many of which are used in leisure.

Early adopters, to the extent conspicuous display motivates them to buy new technology, may also become authorities of sorts in their circles of friends, relatives, and acquaintances who express interest in it. Not only do early adopters enjoy the prestige of possessing an unusual and conceivably valuable, useful good or service, they also get contacted as sources of information about the good or service. Added to the prestige of ownership is that of expert.

There are, however, a number of forces working against this interest in technological conspicuous display. Miguel Helft (2008), writing about early and late adopters of web browsers, suggests that 'millions of users of nearly every type of Internet service and technology, from Netscape to AOL dial-up to old e-mail systems, still prefer to ignore the pitches [from Netscape about discontinuing its technical support for subscribers] and sit still – or at least move at their own pace.' It seems that some later adopters as well as nonadopters are sufficiently pleased with their present technology to resist buying anything new, however highly regarded or even more efficient. Helft found that, though their number is dwindling, more than nine million people in the United States still prefer dial-up service, even while broad band is often available to them at comparable price. Other forces working against the early adoption of new technology include the belief of would-be adopters that new is not necessarily better and the distaste of individuals and businesses for what they define as upgrade 'treadmills' (Helft, 2008). These advances, many later adopters and nonadopters feel, benefit software companies far more than buyers.

Returning to Veblen's thesis, early adopters of technology can parade their unusual, relatively expensive ownership of a valued good or service. Furthermore they enjoy a level of prestige associated with this status and with their expertise on and experience with the novel item. Unlike Veblen's gentlemen of leisure, however, early adopters fail to form a social class. Rather they may be conceived of as enjoying a special personal status in a circle of people interested in a particular technological good or service. Nevertheless this is, most certainly, a modern manifestation of Veblen's theory.

And what about nonadopters, who were just mentioned in passing? It appears that some of them are conspicuously displaying key retro purchases in an effort to signal to their own circles that 'old is good' if not better than that which is new. Hank Stuever (2008) describes the special appeal of the Converse basketball shoe, whose image in certain groups, even after 100 years, is still considered 'cool' and distinctive. Compared with the modern sport shoe, the old simple, rubber-soled, canvas, high-topped model still works nicely for some basketball players as well as some nonplayers, who like the image they convey as they wear them. Likewise with the recreational snowshoer in northern climates who walks on traditional wood-and-leather shoes, while almost everyone else this person meets in a typical outing is moving along on a modern contrivance made of aluminum, stainless steel, or plastic, buoyed by neoprene, nylon, or polypropylene decking.

Competitive compassion

According to the *Cambridge Learner's Dictionary* (2007, p. xviii) competitive compassion is expressed 'when people give money to a charity (= an organization that helps people) because they want to seem kinder than other people *It was competitive compassion which led people to give so much money following the floods*' [emphasis in original]. The floods in question were those caused by the Tsunami of late 2004; it devastated several countries in the Indian Ocean. Misty Harris (2008) maintains that such compassion is 'a trend quickly turning philanthropy into an exercise in self-congratulation.' Writing about Canadians she observes that some wealthy people are parading in Facebook their contributions to charity by, for example, talking about the money they have raised or the number of friends they have managed to rally to their favorite charity or nonprofit organization. Harris says that, 'TV Celebrity Appearance and Oprah's Big Give have transformed philanthropy into a game show, with players competing as teams but being judged as individuals in the fight to be the ultimate altruist.'

The preceding paragraph focused on how competitive is the compassion expressed by individuals. Nevertheless this term, at least in the present-day popular media, is also applied with some frequency to the donations of national governments and, to a lesser extent, those of major, private-sector organizations. Since neither this chapter nor this book delves into conspicuous consumption by social units larger than families, competitive compassion by larger groups will not be examined here. It is, however, a subject well worth exploring, albeit one that would take us too far afield from the scope of this volume.

Though not known by the label, competitive compassion did exist in Veblen's time. Instead he preferred to write about donors of bequests who had 'other motives' in addition to the altruistic one of serving humanity:

> An example of this is seen in the administration of bequests made by public-spirited men for the single purpose (at least ostensibly) of furthering the facility of human life in some particular respect. The objects for which bequests of this class are most frequently made at present are schools, libraries, hospitals, and asylums for the infirm or unfortunate. The avowed purpose of the donor in these cases is the amelioration of human life in the particular respect which is named in the bequest; but it will be found an invariable rule that in the execution of the work not a little of other motives, frequently incompatible with the initial motive, is present and determines the particular disposition eventually made of a good share of the means which have been set apart by the bequest.
>
> (Veblen, 1899, p. 226)

Veblen then elaborated this observation with another: some buildings meant to serve these humanistic interests were instead constructed according to the 'canons of conspicuous waste and predatory exploit.' In other words, publicly visible parts of the structure were designed to impress with their pecuniary excellence rather than to enhance its effectiveness for those it was supposed to serve. Veblen placed this discussion under the rubric of the 'survivals of the non-invidious interest,' showing thereby how this humanitarian interest clashes with the selfish interest of conspicuous display. When it came to public visibility of the bequest, the latter commonly took precedence over the former.

As has been true of much of what has been said in this chapter about modern conspicuous consumption, the competitive compassion of today is far more general than the so-called noninvidious bequests of Veblen's time. The latter were gifts from the men of the non-working leisure class, whereas the former incorporates a much broader group of well-heeled men *and* women who have amassed significant sums, which they then give to charity. This is money gained from ongoing employment and investments bought with its fruits. The 21st century competitively compassionate philanthropists are not, as a group, as wealthy as the leisure man of yore. Many of the first are, by today's standards, *mere* multimillionaires who give away comparatively small amounts of money, as measured across a period of nearly 120 years with inflation of national currency and what it can buy taken into account. And then, too, Veblen's leisure class never had the relatively inexpensive, diffusional advantages of the Internet and the popular television show to broadcast compassion as conspicuous display.

Conspicuous consumption in leisure

It is possible to conceive of conspicuous consumption as the wealthy person's style of what will be discussed more fully in Chapter 5 as 'initiatory consumption.' In such consumption use of the purchased item is immediate or reasonably close to it, as suggested by the word 'means' in the earlier quotation from Veblen: 'conspicuous consumption of valuable goods is a *means* [author's emphasis] of reputability to the gentleman of leisure.' An expensive gift, an extravagant holiday somewhere, a sumptuous feast for friends and associates, must first be bought before they can be enjoyed. The principal difference, and one of great importance, is that these acquisitions have an unusual layer of meaning, namely that of demonstrating the social standing of the buyer as expressed in a flashy display of wealth. In conspicuous consumption the actual leisure appears, most often, to follow the purchase of a good or service such as just described, leading, commonly, to casual or project-based leisure experience.

Nonetheless some types of serious leisure, gained through facilitative consumption (discussed further in Chapter 5), can also be conspicuous. Here the acquired item only sets in motion a set of activities, which when completed, enable the purchaser to use the item in a satisfying or fulfilling leisure experience. For example, someone with the money might purchase season tickets for the best seats in the house, from which to watch the opera or a major league sport. This could be considered conspicuous, initiatory, casual leisure were the buyer a mere consumer of the art or sport, but could be thought of as conspicuous, facilitative serious leisure were this person a buff. *Fans* more or less uncritically consume, for instance, restaurant fare, sports events, or displays of art (concerts, shows, exhibitions) as pure entertainment and sensory stimulation, whereas *buffs* participate in these same situations as more or less knowledgeable, albeit nonprofessional, experts (Stebbins, 2005d, p. 6).[1] The latter have been analyzed as a kind of liberal arts hobbyist (Stebbins, 2007a, pp. 28–29).

Yet, today, by no means all conspicuous consumption may be understood as related in these two ways to leisure. For instance making a highly visible charitable gift is not a leisure activity, nor is purchasing a new form of technology for use at work or managing a nonwork obligation. It would also be logically difficult to classify much of gift giving as leisure. Household objects, often full of special meaning for those who display them (Csikszentmihalyi and Rochberg-Halton, 1981) are often purchased, but are still not understood as having anything to do with leisure (though buying them may be leisure). But, when leisure follows the initiatory conspicuously consumptive act, it is visible and dramatic.

Elite tourism is vividly exemplified in the new interest in space tourism. *Wikipedia* (http://en.wikipedia.org/wiki/Space_tourism, retrieved 11 January 2008) says this 'is a recent phenomenon of tourists paying for spaceflights, primarily for personal satisfaction. Consider three examples of conspicuous project-based leisure. As of 2008, space tourism opportunities are limited and expensive, with only the Russian Space Agency providing transport.' Given the high cost of a trip, nearly all tourists take but one (see www.spacetourismsociety.org/Home.html). More down to earth, so to speak, are the extravagant ocean cruises, where well-heeled guests may conspicuously consume by booking the most expensive state rooms on one of the most celebrated luxury liners. And, for $65,000 (includes a guide and a Sherpa or two) and a good amount of physical conditioning, anyone these days can complete the nontechnical climb up Mt. Everest's north side to gain the coveted title of World Class Mountaineer (Kodas, 2008).

Other examples are mostly related to casual leisure. They include owning an expensive condo as second or third residence in a high-class resort, renting the penthouse suite in a swanky hotel, and holding season tickets for the most expensive *loge* available from which to watch the ballet. Finally, as an instance of conspicuous consumption of services in this century, note that, in England and particularly in Central London, use of butlers by the extremely wealthy as evidence of their elevated status is now said to be rivaling the extent of the practice in Victorian times (Binham, 2008).

Still more down to earth we find numerous instances of modern, conspicuous, leisure-related consumption in circles of people with varied but limited amounts of money for spending on personally validating display. The electronic marketplace offers many an opportunity to look better than one's peers such as by purchasing the latest and most expensive iPod, video camera, or laptop computer. For a period of time Motorola's Razr mobile phone was the rage among teenagers and young adults. It was noted that early adopters sometimes fall into this category, but some later adopters may also be able to shine this way if the item in question is sufficiently costly to dampen widespread sales.

Identity

Conspicuous consumption is, among its many other features we have so far discussed, also very much about identity. Although such consumption has a variety of macro-sociological ramifications as seen, for instance, in the formation of a leisure class and the ranking of economic power in certain primitive societies, everywhere it serves most fundamentally as a foundation for individual differentiation along lines of accumulated wealth. This is a micro-sociological issue. In Veblen's era it was a matter of superior identity among wealthy men on the order of: Mr X is rich, richer than Mr Y and Mr Z, for look at what he can buy that they cannot afford (e.g., a mansion, a lavish dinner and party). Among the Kwakuitl the *potlatch* determined economic rank among families, villages, and tribes, the representative of which was an individual chief whose identity was linked to one or more of these social units.

McCall and Simmons (1978, pp. 62–64) distinguished two types of identity that bear on the present discussion. They observed that most people have several 'personal identities,' or culturally recognized categories of humanity to which they see themselves as belonging. In the realm of conspicuous consumption it is, however, vitally important for individuals who see themselves as members of, say, the leisure elite, to know that others in this select group hold the same view of them. This latter image McCall and Simmons dubbed 'social identity.' Conspicuous consumers believe they have 'arrived' when they can, first, personally identify themselves as rich, and hence as belonging to the leisure class and, then, socially recognize that others identify them likewise.

Identity, both personal and social, as we have considered it to this point is based on placing well in the competition to display the most wealth. Here we have *competitive consumer identity*, of which Veblen's discussion of noninvidious interests is a fine example. It appears that the most common understanding of conspicuous consumption is anchored in this genre of identity.

But, still, not all consumer identities are nurtured in competition. Indeed casual observation suggests that most are not. That is, people often buy goods and services to show their identity as members of a particular group or category, to which I will refer here as *group consumer identity*. Here there is no competition, only a desire to buy what is needed to prove to people of a particular group or category that the consumer is a bona fide member of it. The purchased good or service may not even be expensive, but all or nearly all members are identified by it (usually along with several other indicators).

Some group consumer identity symbols may only be purchased from special sources exclusively available to authenticated members. A card, pin, badge, jacket, plaque, or certificate may only be accessible to those who pay a fee for membership in the group or organization or even directly buy such an item. True some club memberships come only at a hefty price, but the identity generated by paying it may well be for members who can easily afford it more a matter of consumer group identity than of competitive consumer identity. For them it is not how much they paid for this identity, but that they were allowed (invited, certified) to buy it.

Another class of consumer identity symbols is, by contrast, freely available on the open market. For example, anyone with sufficient money can buy a nose ring and have it placed, expressly to demonstrate to all concerned the wearer's membership in a distinctive group of youth. Certain kinds of apparel (hats, caps, shoes, pants, etc.) may serve a similar purpose. The same holds for particular kinds of equipment such as skis, musical instruments, running shoes, and golf clubs. Informed skiers, musicians, runners, and golfers recognize like-minded participants by the level of sophistication of the skis, instruments, and so forth that the latter use in their work or leisure. In this regard Tanner, Asbridge, and Wortley (2008) show how musical styles identify adolescents as belonging to certain peer groups, and conversely, how members of these groups identify with the style associated with them. The style of music in question is sometimes consumed through purchases of, for example, a CD, concert ticket, or Internet downloading fee.

Recently Toyota created a variation on this theme by offering a range of optional decorations on its automotive model, 'Scion.'

> TOYOTA likes to think of its quirky, boxy Scion as a 21st-century chariot of the soul – not just an affordable car, but also a unique expression of the young, hip person who Toyota hopes is driving it.
>
> Now Toyota's Scion enthusiasts will have even more 'me time': a marketing campaign with an underground vibe that is intended to show just how much their chosen transportation reflects their personality.
>
> With an eye to the social networking ethos that has made Facebook and MySpace wildly popular, Toyota will let Scion owners design their own personal 'coat of arms' online, a piece of owner-generated art that is meant to reflect their job, hobbies and – um, O.K. – karma.
>
> In making their personalized crests, Scion owners can choose from among hundreds of symbols, all designed by a professional graffiti artist. The symbols range from an eagle, a jester, a king's crown and a worker's fist to Japanese anime-style flowers, a three-person family and a yin-yang circle. Customers can download their designs and have them made into window decals or take them to an auto airbrushing shop to have them professionally painted onto their cars.
>
> The Scion is an economy car aimed at younger, stylish drivers, and the design Website, scionspeak.com, is free. But Scion enthusiasts must pay for the auto shop renderings of their design, an indulgence that can cost thousands of dollars.
>
> The campaign, called Scion Speak, was created by StrawberryFrog, an advertising and marketing agency based in New York and Amsterdam that is known for its quirkiness and for representing new or hipster brands. The agency spent six months last year escorting a graffiti artist, Tristan Eaton, around New York, Los Angeles and other cities to talk to Scion owners about their lifestyles. Based on those conversations, Mr. Eaton designed the symbols.
>
> 'These guys love to personalize their cars, and we give them a tool to do that,' said Kevin McKeon, the executive creative director of StrawberryFrog in New York
>
> (Browning, 2008)

Tourism

The world of tourism is rich in opportunities for conspicuous consumption and the creation of identities that flow from a person's travels. John Urry observed (1994, p. 235) that the tourist identity is especially prominent in the postmodern age. He argued that: 'identity is formed through consumption and play. It is argued that people's social identities are increasingly formed not through work, whether in the factory or the home, but through their patterns of consumption of goods, services and signs.' According to Urry, tourism in the postmodern age has become a main pattern of consumption.

Nevertheless, he failed to note that tourists' identities may vary by type of tourism. Thus, considering cultural tourism a form of serious leisure and mass tourism a form of casual leisure (Stebbins, 1997b), the first, being far more exclusive than the second, should also offer far greater scope for distinctive self-identification. Furthermore, if general, or mass, tourism is a source of distinctive self-identification, then the more exclusive, profound cultural tourism must be a source of valued identity extraordinaire.

Why should cultural tourism generate, especially, distinctive identities? The answer to this question lies in the condition that all serious leisure, this type of tourism included, roots in the six distinguishing qualities separating it from casual leisure (discussed in Chapter 1). More particularly cultural tourism is a liberal arts hobby. Compared with mass tourism the identity base of its cultural counterpart is substantially different, giving those who go in for the latter an identity of far greater depth and complexity than available to those who go in for the former. Mass tourism, by its very definition, is socially, financially, and geographically accessible to great numbers of people, as seen, for example, in much of guided tourism, camper tourism, and psycho-centric tourism (Plog, 1991, pp. 62–64). In contrast the objects attracting cultural tourists are socially and psychologically much less accessible as a rule, since they require these tourists to develop certain tastes (e.g., in art, food, music, or architecture), acquire certain kinds of knowledge (e.g., a foreign language, the history of a region or country), or polish particular social skills (e.g., how to talk with locals, how to act according to local norms). By the way, many of these personal acquisitions needed as background for the trip must themselves be purchased, as in language lessons, guide books, and possibly vaccinations against infectious diseases. This is consumption, to be sure, and directly related to leisure, but it is for the most part not of the conspicuous variety.

Yet, even in cultural tourism, mass tourist type purchasing sometimes enters the picture, albeit with a special twist. Maurice Kane and Robyn Zink (2004) found that package adventure tourists displayed their identity as such by wearing certain visible kinds of apparel (e.g., shoes, shirts, jackets). The intent was to call attention to their seriousness in touring, to themselves as adventure tourists rather than ordinary mass tourists.

Apparel

In part, because much of it is so visible, apparel as a vehicle for conspicuous consumption and identity display necessarily becomes a main subject in any discussion of these two processes. It should be no surprise, then, that the conspicuous consumption of clothing caught Veblen's eye. He devoted an entire chapter to the matter (Veblen, 1899, chapter 7).

> For this purpose no line of consumption affords a more apt illustration than expenditure on dress. It is especially the rule of conspicuous waste of goods that finds expression in dress, although the other, related principles of pecuniary repute are also exemplified in the same contrivances. Other methods of putting one's pecuniary standing in evidence serve their end effectually, and other methods are in vogue always and everywhere; but expenditure on dress has this advantage over most other methods, that our apparel is always in evidence and affords an indication of our pecuniary standing to all observers at the first glance. It is also true that admitted expenditure for display is more obviously present, and is, perhaps, more universally practiced in the matter of dress than in any other line of consumption.
>
> (Veblen, 1899, pp. 118–119)

Moreover, says Veblen, the greater part of consumption of apparel is for appearances rather than for protecting the person against the elements. Indeed, then as now, some people even go 'ill clad in order to appear well dressed.' This is one of a small number of sections in Veblen's book where women figure prominently in his treatment of conspicuous consumption, the bulk of it being centered on men.

Gregory Stone (1962, pp. 101–104) shows, in concrete terms, how identification by apparel unfolds through four stages of validation of the self of the wearer. He says that people clothed in particular items of apparel 'announce' their personal identity this way, evoking thereby their 'placement' by an audience, or those present in the same situation. In McCall and Simmons's terms the second socially identify the first. This announcement is simultaneously a 'show' by the wearer, which encourages the audience to 'appraise' what they see. This is the 'value' stage of validation. Next comes an emotional 'expression' by the wearer to the review of his or her apparel as it has progressed to this point and in reaction to the audience's 'appreciation.' Stone referred to this the 'mood' stage of validation. In the final stage, called 'attitude,' the wearer, in effect, makes a 'proposal.' It is comprised of a subtle set of signs emanating from that person's behavior in past and present, which is scrutinized by the audience for its future consistency with the apparel chosen. The audience, on the basis of what they see, 'anticipates' this consistency or lack thereof.

Stone's validation model elaborates further Veblen's theory of conspicuous consumption of apparel. The model shows the interpersonal processes at work when audiences appraise someone's dress. It looks into the emotional side of consumption, as when the appreciation validates the wearer's claims and when it fails to do this. For, if consumptive identification according to the first two types is incommensurate – that is, if social identity fails to corroborate personal identity – the victim in this situation is sure to be upset in some way. Finally Stone's model, through the idea of attitude, introduces consideration of how audiences will possibly treat the wearer in future, as based on past and present sartorial evidence. In other words, do the clothes make the man, in the sense that he is now regarded as a true member of the leisure class (or, today, other exclusive group or category)?

This book is being written at a time in the economic cycle when the high prices of fashionable clothing and accessories at prominent New York retailers are in free fall. According to Guy Trebay (2008) Saks, for example, cut it prices on such goods for the 2008 Thanksgiving Weekend by as much as 70 percent. So, a Valentino evening dress originally tagged at $2950 (US) was on sale at $885 (US). What then of shoppers who announce their identity as people who have the money to buy such items at their original price, even while they paid 70 percent less for them? Will they be discovered during the proposal stage of consumption as social-class imposters? In other words, will their conspicuous display fail to generate the level of placement they hoped to achieve?

The global postmodern tribe

While it is true that, among certain primitive peoples, tribes function as an important form of local social organization, these kinship groups are really only a metaphor for Michel Maffesoli (1996). Instead he transforms this relatively narrow anthropological concept into one much broader and sociological, which identifies and describes a postmodern phenomenon spanning national borders. In this regard, he observes that mass culture has disintegrated, leaving in its wake a diversity of 'tribes.' These tribes are fragmented groupings left over from the preceding era of rampant mass consumption, groupings recognized today by their unique tastes, lifestyles, and form of social organization.

Modern tribal groupings exist for the pleasure of their members to share the warmth of being together, socializing with each other, sometimes seeing and touching each other, and so on – a highly emotional process. In this they are both participants and observers, as exemplified by in-group hairstyles, bodily modifications, and items of apparel. This produces a sort of solidarity among members not unlike that found in the different religions and primitive tribes. Moreover, being together under these conditions can lead to a kind of spontaneous creativity that gives rise to widely varied, new cultural forms having appeal for great masses of people. Some of these new tribes are negative, if not deviant, seen in, for instance, those fostering local racism and ostracism (e.g., the skinheads). Whatever their moral basis all tribes beget distinctive 'lifestyles,' although in casual leisure they are much less complicated than those springing from serious leisure pursuits. The postmodern world is, among other things, a 'multiplicity of lifestyles – a kind of multiculturalism.'

Maffesoli argues further that today's tribes serve as antidotes to the dominant individualism of our time, since individual identity is submerged in such groups. They are not, however, without ideals. Rather he observes that 'it would perhaps be better to note that they have no vision of what should constitute the absolutes of a society. Each group has its own absolute' (Maffesoli, 1996, pp. 88–89). There is, moreover, a secret sharing among members of the emotions and experiences unique to their tribe, a process that reinforces close group ties and distinguishes insiders from outsiders. As for context, postmodern tribalism must be seen as a product of the massified metropolis, a distinctively urban phenomenon.

Much of postmodern tribalization has taken place in the sphere of leisure, where it has given birth to a small number of interest-based, serious leisure tribes and a considerably larger number of taste-based, casual leisure tribes. Although Maffesoli (1996) fails to recognize these two leisure forms, Rob Shields (1996, p. xi) briefly mentions them in his Foreword to the English translation of the former's book. 'Typical examples of *tribus* [tribes],' says Shields, 'are not only fashion victims, or youth subcultures. This term can be extended to interest-based collectivities: hobbyists; sports enthusiasts; and more important, environmental movements, user-groups of state services and consumer lobbies.'

In addition, although there is no gainsaying that serious leisure and its enthusiasts are marginal in ways set out earlier in this book, this condition in postmodern times seems only to enhance their sense of tribal belonging. When the larger community sees these tribes as quaint, eccentric, or simply different, solidarity among members is strengthened in significant measure. And this pertains even though all must live with this unfavorable image. Meanwhile, in some forms of tribal serious leisure, a small number of leisure organizations provide their members with socially visible rallying points for individualized leisure identities as well as outlets for the central life interest they share (Stebbins, 2002, Chapter 5). Most often this organization is a club, which nevertheless serves as an important axis for the lifestyle enjoyed by enthusiasts pursuing the associated serious leisure activity.

The casual leisure tribes are unable to offer this more complex level of organizational belonging, in that they rarely, if ever, become even this formally organized. These tribes retain too much of their former character as consumer masses to serve as the seedbed for formal leisure groups and organizations. There is, of course, a true sense of belonging that comes with sharing private symbols with other members of the same mass (Maffesoli, 1996, pp. 76–77, 96–100). Yet, the feeling of solidarity that comes with belonging to, for example, a small group or grassroots association is commonly missing in taste-based tribes.

Taste-based tribes are especially popular among contemporary youth, being a main trend these days in this age category and favored over earlier tendencies to join established groups. Ken Roberts, in writing about their tribes, observes that

> the groups of young people (and adults) who become players in, or fans of spectator sports teams, and who attend 'raves' and similar scenes where their drugs of choice are available and their preferred types of music are played, can experience intense camaraderie.... Much of the appeal of these occasions is that they are incredibly social. Individuals find that they are accepted and experience a sense of belonging. None of this is completely new. The change over time has been that the young people who play together nowadays have rarely grown up together and attended the same local schools. Their sole bond is likely to be leisure taste or activity. Yet being part of these scenes can be extremely important to those involved.
>
> (Roberts, 1999, p. 9)

Taste-based tribes are not, however, the exclusive social domain of young people (Stebbins, 2002, pp. 65–69).

Many of today's tribes exemplify well the group consumer identity, in service of which members often buy goods and services to prove their identity as belonging to one of them. Andrew Bennett puts in historical context the growth of this tendency.

> In my view, then, the process of tribalism identified by Maffesoli is tied inherently to the origins of mass consumerism during the immediate post-Second World War period and has been gathering momentum every since. That it should become acutely manifest in the closing years of the twentieth century has rather more to do with the sheer range of consumer choices which now exist than with the onset of a postmodernist age and attendant postmodern sensibilities.
>
> (Bennett, 2003, p. 154)

We return to the subject of tribes in Chapter 6.

Conspicuous waste

Veblen was also greatly concerned about the waste generated through conspicuous consumption by the leisure class. To be sure such people sometimes bought goods and services to meet basic necessities, but then, too, they made numerous purchases mainly to flaunt their economic and social superiority. There was thus a moral issue underlying the practice, an issue that cried out for both examination and exposure. Veblen felt society needed to be aware of its deleterious consequences.

In Veblen's day the waste in question was evident in the consumption of alcohol and tobacco, banquet leftovers thrown into the garbage, palatial homes with unnecessarily large rooms, decorations made of materials that deplete the environment, and the list goes on. Chris Rojek (2005, pp. 90–91) observes that such waste continues unabated, this list having become even longer. Today we waste by, for example, preparing or buying fast food, taking recreational drugs, relying on petrol-based transportation, and applying spray perfumes. Moreover such waste can often be traced to conspicuous display intentionally carried out during leisure activities.

A good deal of this leisure-related waste, Rojek says, is associated with casual leisure. He examined the modern working class, which has developed its own way of conspicuously consuming. There he found a 'celebration of autonomy, control and resistance' to the inequalities they are forced to endure (Rojek, 2005, pp. 149–150). In various 'ceremonies of excess' many individuals and groups in this stratum excessively drink alcohol, take narcotics, and consume tobacco. For them it is fashionable to use blue language and gestures and engage in aggressive behavior and uninhibited sexuality. Much of this is associated with intense camaraderie. Through such practices participants in this culture develop a distinctive consumer identity, offering thereby a stark contrast with the practices and emergent identities of the leisure class. In both groups there is significant waste.

But not all waste is related to casual leisure. Some of it is related to serious leisure, although here waste, consumption, and leisure may have very different relationship. For instance, some people find their leisure in criticizing this culture of excess, doing so as career volunteers who work for social movements whose goals are to reduce smoking, protect the environment, or enhance health through good nutrition and beneficial exercise. Leisure identities based on such activity are not formed through emulation (of dominant roles and personalities), but rather through a sense of opposition to correcting what is seen as a crucial fault in mainstream society. This leisure is not based on wasteful consumption, but instead, is experienced while trying to reduce such consumption.

During the dialectical interplay over the years between wasteful behavior and the counter-waste social movements, we have witnessed a slow but steady emergence of the simple living movement, a formation (covered more fully in Chapter 6) that is decidedly anticonspicuous consumption. Briefly put simple living is the process of reducing the superficial aspects of our existence, so as to allow more time and energy for developing those aspects we prize as worthwhile. In fact aspects of the simple life, as voluntary simplifiers see things, are anticonsumerist, against all purchases save those meeting basic needs. Some aspects may also be 'non-adoptionist,' in that some nonadopters not only like certain retro goods but also want to economize on their purchases.

Given that, in pursuing either serious leisure or project-based leisure, participants make many contributions to the community and that these two forms offer two avenues for realizing human potential, it is reasonable to interpret participation in such leisure as consistent with the principles of simple living. Note, however, that since its adherents also espouse many other principles, this way of living is by no means identical with serious leisure. Nevertheless, the two do share the common ground of encouraging and fostering self-fulfillment through realizing individual human potential while contributing to the well-being of the wider community. For the typical, true believer in simple living, this often means paring back work activity as well, where such activity generates more money than needed for a simple lifestyle and uses up time that could be spent in self-fulfilling leisure.[2]

Conclusions

There is a tendency in university circles, evidenced by its sparse coverage in modern times, to treat conspicuous consumption as strictly of historical concern, an intriguing, scholarly analysis of late 19th century social class in the United States. True, today's graduate students in theoretically oriented courses in leisure studies, sociology, and economics may occasionally be asked to read Veblen's little book. And it makes good reading, wonderfully written and brimming with interesting passages as it is. But this assignment seems more a mechanism for establishing an appreciation of the intellectual history of these disciplines than one for providing analytic tools for understanding present processes and practices.

Be that as it may conspicuous consumption, conspicuous display, and the waste associated with both are far more relevant for contemporary Western society than they were for the society about which Veblen wrote. First, remember that these three processes have a long history, partial proof of which lies in the ways certain primitive groups have enacted them. Veblen's study is no isolated analysis of a culturally and historically unique phenomenon. Second, conspicuous consumption, display, and attendant waste are far more widespread in modern society than in Veblen's society or in the societies of the primitive groups. We may talk about a leisure class in the West in the 21st century, but its boundaries are much vaguer than heretofore, and as pointed out, many people consume and waste conspicuously who have full-time employment and who may be inspired, if not driven, by the work ethic. Third, the waste that flows from such practices is a growing social and environmental concern. Simple living is, among other things, a prescription for bringing this harmful tendency under control. But, today, the stakes are higher than in Veblen's time (or at least we now recognize these stakes). The planet, many scientists believe, is warming at an alarming rate with grave consequences in the offing. Controlling waste, both conspicuous and otherwise, has become critical. Simple living is the practice by which this goal can, in part, be reached.

All this elevates substantially the importance of leisure in modern society. For this is a main domain of life where conspicuous consumption, display, and waste frequently occur, and we may thank Veblen for giving us some conceptual tools with which to see this. Moreover it is substantially through leisure that simple living is effected as a lifestyle. Leisure may appear to some people to be little more than hedonism, trivial casual activities hardly worth scholarly consideration. The broader picture sketched in this conclusion, and more generally in this chapter, suggests a dramatically different appreciation of what we do in free time.

3
Consumption and Leisure in Context

This chapter looks further into the personal and social conditions in which the consumptive leisure experience of today takes place. The personal conditions include health, wealth, marital status, level of education, taste and talent for a given product and accompanying activity, knowledge of the product and the activity, and the like. Numbering among the social conditions are advertising and promotion, historical forces, type of government, local and national culture, gender stereotype, ethnicity (including religion, race, nationality), consumer advocacy groups and services, geographic location, and others. Still, the scholarly literature in this area is not nearly as broad ranging as these lists. In fact, only a small, though perceptive and sometimes controversial, set of thinkers have addressed themselves to a select set of personal and social conditions that they considered crucial to understanding consumption as they observed it at the time they wrote.

The contributions of these men and women are covered here in approximate chronological order, with special attention given to their import for consumption and leisure. I make no attempt to provide representative coverage of the literary corpus of each. Rather this book is about leisure and consumption, and the works of these scholars are examined only to the extent that they relate to these two aspects of modern life.

Karl Marx

Consumption as understood and practiced in the modern Western world got its start in the Industrial Revolution, the birthplace of widespread monetary exchange, mass-produced goods, and desire and wherewithal to buy such goods in the marketplace, goods that went beyond meeting basic needs. Marx (1818–1883), who, though born in Germany, spent much of his life in England, and was therefore well positioned to observe the effects of this great historical transformation that had unfolded in Britain during the late 18th and early 19th centuries. Capitalism was the ideological engine that inspired and directed the Industrial Revolution, and its effects were not universally beneficial for all concerned. Marx's intellectual career centered on various negative consequences of capitalist-style industrialization, notable among them, the exploitation of workers, the dialectical forces leading to revolutionary change, and the role of religion as handmaiden to these processes.

From one angle consumption figures in Marx's thought mainly as a middle link in the chain of production–consumption–capitalist wealth. That is, industrialized workers 'toil' at their jobs to produce goods, which they are also expected to buy with the money they earn. These sales ring up profits for the owners of the enterprises – the capitalists – who employ these workers. In this arrangement workers must be paid well enough to buy goods beyond those required for subsistence, but not so well as to cause owners' profits to shrink appreciably.

But from another angle consumption, as the term is used in this book, was conceived of by Marx as the final step in a process which begins with production (Marx, 1977, pp. 349–350). In this conceptualization he sketched the basic process a product follows, starting with its production, proceeding through its distribution and exchange in the middle, and culminating in its consumption. Consumption meets individual needs. Distribution takes place according to social laws; it determines the proportion of a society's members having access to a given product. Through exchange these members purchase the product with an eye to satisfying certain needs. Distribution is determined by society, exchange by individuals and their needs.

Marx's interests lay chiefly in the nature and consequences of the social conditions underlying consumption. He said little about whether the goods bought were for leisure and how such purchases might relate to the use of free time. Be that as it may, Marx lived at a time when most people worked long days, had therefore precious few discretionary hours, and mostly purchased things that filled basic needs. This was the sort of consumption at the end of the production–consumption chain on which Marx centered his attention. Marx wrote about consumption (as it filled basic needs) in the context of economic activity at a point in history when the Western societies were industrializing.

Georg Simmel

Georg Simmel (1858–1918), born and educated in Germany, was one of the first generation of the world's sociologists. Typical of the social scientifically oriented scholars of this period, he was trained in philosophy and history (he also studied social psychology). Indeed Simmel's essay on fashion, that part of his *œuvre* of interest in this book, has sometimes been translated from the German as the philosophy of fashion. As for Marx's influence, there appears to be little evidence of it in Simmel's observations in this essay, notwithstanding the considerable stature of the first in European thought at the time.

Simmel argued that the very character of fashion requires it be exercised at only one time by a portion of a particular group, with the great majority of that group being merely on the road to adopting it. In the language of the preceding chapter, the minority may be seen as early adopters. But, says Simmel,

> as soon as an example has been universally adopted, that is, as soon as anything that was originally done only by a few has really come to be practiced by all – as is the case in certain portions of our apparel and in various forms of social conduct – we no longer speak of fashion. As fashion spreads, it gradually goes to its doom. The distinctiveness which in the early stages of a set fashion assures for it a certain distribution is destroyed as the fashion spreads, and as this element wanes, the fashion also is bound to die.
>
> (Simmel, 1971, p. 302)

By dint of this peculiar interplay between the tendency toward universal acceptance and the destruction of the very purpose of fashion caused by this general adoption, the distinctive attraction of fashion is its limitation. In other words, there exists an 'attraction of a simultaneous beginning and end, the charm of novelty coupled to that of transitoriness' (Simmel, 1971, p. 302).

One of the attractions of something fashionable for early adopters is this contradictory duality. Fashionable people know that their social distinctiveness generated by the unusualness of what they wear or do (for Simmel, patterns of behavior may also be fashionable) will not ultimately be theirs exclusively. Therefore 'few phenomena of social life possess such a pointed curve of consciousness as does fashion' (Simmel, 1971, p. 303). This quality, among its early adopters, only makes a given fashion that is much more appealing.

Simmel failed to address himself to the implications of this thesis for leisure. Using the framework set out in Chapter 5 of this book, it is nonetheless possible to classify the purchases of fashionable goods and services as 'initiatory consumption.' In other words, the actual leisure appears, most often, to follow such purchases, most commonly leading to a casual or project-based leisure experience. In consumption of this sort use of the purchased item is either immediate or reasonably close to it. Moreover fashionable purchases are by their very nature occasions for conspicuous consumption. As for intellectual context Simmel's ideas on fashion may be categorized as sociological and social psychological.

Henri Lefevbre

Henri Lefevbre (1901–1991) was a widely known French Marxist intellectual, who wrote on a fair range of subjects, among them, contemporary consumptive practices. He used a number of Marx's ideas in combination with his own existentialist, phenomenological, and structuralist observations to develop a critical Marxian perspective on this aspect of social life. His work on leisure and consumption is found mainly in his book *Critique of Everyday Life* (Lefevbre, 1991, French edition published in 1947).

Lefevbre stressed the importance of everyday life, as the sphere where all activities, routine and otherwise, take place, contrary to the established philosophic position of the day that this sphere is trivial. Instead Lefevbre contended that it is in everyday life where the self is formed and the creative potential of each person is realized. This stance was meant as a critique of Marx and the structuralists, both of whom denied the role in social change that individuals might play. More particularly, Lefevbre was concerned with positive human freedom, freedom to 'do' rather than freedom 'from,' for example, tyranny or excessive control by the State.

> Liberated from sordid necessity, needs per se are becoming suffused with reason, social life, joy and happiness. Especially in our era, the condition which restricted creative leisure and spiritual activities to the oppressors has disappeared. It is a complex dialectic: needs are becoming more extensive, more numerous, but because the productive forces are broadening, this extension of needs may imply their humanization, a reduction in the number of hours worked to satisfy immediate needs, a reduction of the time

spent at work generally; a universalization both of wealth and of leisure.

(Lefevbre, 1991, p. 175)

In the final analysis, however, Lefevbre regarded leisure as a villain. He considered hobbies, film, art, sport, and the like – generally identified as serious leisure in the present book – as distractions from and compensations for work. As such these activities were held, under capitalism, to be alienating. 'We work to earn our leisure, and leisure only has one meaning: to get away from work. A vicious circle' (Lefevbre, 1991, p. 40). The chief mechanism behind such alienation in modern society is creation by the advertising industry of leisure wants, wants capable of being met only through buying particular goods or services. This is a false world, a facsimile of real pleasure and fulfillment. This 'illusory reverse image' is served up today in, for instance, most film, press, theater, and music hall.

Lefevbre brought a unique perspective to the determinism/free will debate. He stressed the importance of human agency in leisure (freedom to do), on the one hand, while underscoring the constraining alienation of modern leisure as promoted through advertising and commodification, on the other. In this fashion he rode the fence separating the pessimism of the mass leisure theorists and the more positive stance of thinkers who have written about the consumer society (see Chapter 1). The context of leisure and consumption provided by Lefevbre is fundamentally sociological, notwithstanding his schooling in philosophy and the philosophical character of much of his writing.

Jean Baudrillard

Baudrillard (1929–2007), a major 20th-century French philosopher, sociologist, and cultural theorist, drew much of his early intellectual inspiration from Marx. Nevertheless the former's ideas on consumption are vastly different, in fact a reaction to Marx's position on production and consumption. On matters of consumption Baudrillard takes a philosophic/sociological stance (Baudrillard, 1981, pp. 63–68). There is a 'logic' to consumption, he argues. Or, more precisely, there are four logics: One, there is a functional logic of use value, or a logic of practical operations, of utility. Two, there is an economic logic of exchange value, or a logic of equivalence, of the market. Three, there is a logic of symbolic exchange, or a logic of ambivalence, of the gift (Baudrillard's indebtedness to Mauss is evident here). Four, there is a logic of sign value, or a logic of difference, of status. Thus an object of consumption assumes the status of (1) an instrument, (2) a commodity, (3) a symbol, or (4) a sign.

Baudrillard provides several examples of the logic of consumption. Consider the one of buying a house or renting an apartment – in general, acquiring lodging. It has practical value for most people, a functional logic. It also has exchange value, in that it is probably to be bought or rented in the housing or apartment market, an economic logic. For some people their house has symbolic value, since, for instance, they inherited it, acquired it as a gift, or won it in a lottery. Finally, most people's houses and apartments in Western society have sign value, demonstrating a logic of difference. Here a person's dwelling is a sign of wealth or its lack, of taste, of social class, and the like.

Alan Warde places this part of Baudrillard's work in the field of cultural studies, a discipline that has tended to challenge the negative conceptualizations of consumption treated of in Chapter 1 under the heading of 'mass consumption.' According to Baudrillard,

> *consumer culture* provided entertainment and stimulation, people engaged with manufactured cultural artifacts in active and creative fashion, and some groups employed the items of mass consumption to subvert dominant values and norms. Baudrillard's (1998) insistence on seeing consumption as primarily a system of signs, rather than a source of use-value, was a notable contribution to the emergent understanding of postmodern culture as visual, transient, ephemeral, and playful, increasingly a means of expression of personal and collective identity.
>
> (Warde, 2005, p. 58)

Contextually speaking Baudrillard, in his writings, often looked at consumption through the lens of popular culture. The playful aspect of consumption that he observed hints at leisure, without however, going further into the ways that playfulness manifests itself in the vast world of free-time activity. He also brought a semiological perspective to bear on his interests. Indeed his early work was one of the first to utilize semiology to analyze how objects are encoded with a system of signs and meanings as these operate in both the modern mass media and the wider consumer society. Most broadly Baudrillard contextualizes the area of leisure and consumption in a unique way, by combining semiological studies, Marxian political economy, and sociology of the consumer society at the level of objects and signs in everyday life (Kellner, 2003).

Of Baudrillard's early books those titled *For a Critique of the Political Economy of the Sign* and *The Consumer Society* bear most directly on the present discussion. The first of these was covered in the early paragraphs of this section. In the second Baudrillard focuses primarily on consumerism, extending the analytic scheme set out in the first wherein he sees different objects as consumed according to different logics. This was Baudrillard's point of departure from Marx, since the latter was chiefly interested in production as the mainspring of capitalist society whereas for the former this mainspring was consumption. This is a fair criticism of Marx, in the sense that his ideas, by mid-20th century, needed revision in face of a level of mass consumption heretofore unknown.

But, for Baudrillard, consumption is more important than production, because the 'ideological genesis of needs' (Baudrillard, 1981, pp. 63–68) leads to production of goods designed to meet those needs. In the four-fold model set out above, Marx saw the production process as beginning in the functional and economic logics, whereas Baudrillard stressed the logics of symbol and sign as the starting points from which to explain consumption. In this latter conception there is scope for creativity and inventiveness – playfulness – as people look around for ways (products) to fulfill needs of various kinds. These needs are constructed, interpreted as such by 'needy' people. Even basic, innate needs are subject to this process, as in thirst (drink water, beer, a soft drink?), hunger (eat a sandwich, candy bar, bowl of soup?), or sexual satisfaction (stimulated by pornographic film, autoerotic device, chemical aphrodisiac?). Whatever the need practical, need-meeting products are then produced and subsequently marketed.

In the meantime what we buy to meet our needs, basic or otherwise, 'gives off' to all who care to observe (this is the terminology of Erving Goffman, who also influenced Baudrillard), signs about us as individuals. Thus, drinking beer to slake our thirst, at the very least, signals that we are not teetotalers. The brand consumed may give off further identifying information, as may the situation in which it is consumed (e.g., bar, picnic, private home, fishing trip) and with whom this is being done (e.g., spouse, coworkers, same-sex friends, extended family). There is no doubt that the modern design, production, and marketing of many a consumer good and service is substantially based on these personal interests, to the extent they are known by manufacturers and shared by a targeted group large enough to generate significant profit.

In *The Consumer Society* Baudrillard extended his individualist stance by championing the 'subculture of non-violence' and the 'refusal' of social convention and conspicuous consumption (yes, he was impressed by Veblen's analysis). He saw these two as ways of challenging the wide variety of conformist thought and behavior, which had become rampant during the second half of the 20th century (Baudrillard, 1998, pp. 179–185). Moreover, with refusal, the seeds are sown for radical social change.

But with refusal Baudrillard also abandoned the positive conceptualization of consumption observed by Warde to embrace the more cynical, negative outlook of the critics of mass society. In the final chapter of *The Consumer* Society, he predicted 'violent eruptions and sudden disintegration which will come, just as unforeseeably and as certainly as May 68, to wreck this white mass [consumption]' (Baudrillard, 1998, p. 196). Drawing again on Marx he saw alienation becoming so ingrained that it will become unshakable, rooted as it would be in the very structure of the market society. This will happen because everything can now be bought and sold. Ours, today, is a commodified society – everything is a commodity – consequently alienation will be both inevitable and profound. What is worse we are in for 'the end of transcendence' (Marcuse's term). In this state of mind people will be incapable of perceiving their own needs as well as the alternatives they might pursue for a different, largely noncommodified, way of life.

In Chapters 4 and 5, of the present book, I will argue that this pessimistic outlook lacks both empirical and theoretical support. More particularly leisure and the consumption of goods and services and their interrelationship are far more complex than acknowledged in these sweeping claims. The claims betray a deeply inadequate understanding of modern leisure as a vast set of activities and of how consumption relates to these activities.

Michel de Certeau

Baudrillard broke with Marx when the first began to emphasize the scope people have for creativity and inventiveness – playfulness – as they look around for ways (products) to fill their diverse needs. De Certeau (1925–1986) shared this orientation, which is most fully expressed in *The Practice of Everyday Life* (de Certeau, 1984). The French original from which this book was translated – *L'invention du quotidien. Vol. 1, Arts de faire* – was published in 1980, some years after Baudrillard's *Pour une critique de l'économie politique du signe*, published in 1972 (The English translation was published in 1981 as *For a Critique of the Political Economy of the Sign*). Thus, although de Certeau was undoubtedly aware of Baudrillard's ideas on this subject, the first clearly had his own understanding of it.

De Certeau referred to consumers as 'users,' whom he saw as making unpredictable purchasing decisions in contemporary society dominated as it is by consumer interests. These decisions take place in everyday life. Here they are creative, even playful, inventive, responses to meeting ordinary needs. Although de Certeau stopped short of endorsing the practice of refusal theme promoted by Baudrillard, the first did look on consumption as a routine way of either bypassing or undermining the reigning commercial order of modern times, an internal force fueled by the common man. In this sense consumers (users) become active producers, but what they produce is the everyday life of each person. Each forms, through consumption, a personal world within which to live on a daily basis.

This said de Certeau saw consumer creativity and inventiveness being expressed within a larger context of dominance, within an environment of power relations. The consumer is by no means fully sovereign. In recognition of this principle de Certeau distinguished 'strategies' and 'tactics,' where institutions are analyzed as 'strategic' and ordinary people as 'tactical.' A strategy is either a social formation – an institution, organization, government, commercial entity – or an individual recognized in society as having power. A strategy is part of the dominant order. Its products vary; they might be laws (government), rituals (religions), commercial goods and services (stores), literature (newspapers, magazines), and the like. Strategies are relatively inflexible; in a word they are institutionalized.

Users, or consumers, when they must deal with strategies, often find they fail to fit the latter's conception of who they are. That is the latter commonly conceive of the former in generalized terms, as broad classes of people, clients, for whom certain products apply. The most direct contact with users a strategy may have is through some sort of sampling procedure such as a survey, case study, or focus group.

This situation encourages the use of tactics on the part of consumers. Tactics are not institutionalized. Nor are they visible as strategies are. Rather they depend on time, in that would-be consumers must be always on the watch for occasions to use them. '[A tactic] must constantly manipulate events in order to turn them into "opportunities." The weak must constantly turn to their own ends forces alien to them' (de Certeau, 1984, pp. xi–xxii, 205–207).

De Certeau went on to explain, in concrete terms, how tactical behavior works in everyday life.

Many everyday practices (talking, reading, moving about, shopping, cooking, etc.) are tactical in character. And so are, more generally, many 'ways of operating': victories of the 'weak' over the 'strong' (whether the strength be that of powerful people or the violence of things of an imposed order, etc.), clever tricks, knowing how to get away with things, hunter's cunning, maneuvers, polymorphic simulations, joyful discoveries.

(de Certeau, 1984, pp. xi–xxii, 205–207)

Among de Certeau's many examples of the use of tactics is the one of reading; it has the quality of 'silent production.' The eye ranges across the printed text, picking up what it wants by leaping over some words and dwelling on others. By improvising meanings and harmonizing other meanings with the reader's perspective on the printed content, silent production is effected. Be the text a book, magazine article, or advertising brochure, tactical readers treat the occasion as an opportunity to understand what they read as they wish. Here there is inventiveness, with which these readers undermine, mostly unintentionally or at least unconsciously, a part of the established order.

Obviously de Certeau's ideas, as just presented, apply to a far greater swath of society than leisure and consumption, the focus of the present book. Yet the commercial world of selling goods and services is strategic, and consumers do use tactics in huge variety to adapt it to their needs. This applies to leisure, but it applies equally to work and non-work obligation. In brief, there is nothing about de Certeau's thought that particularly bears on leisure and consumption. Nonetheless it will be helpful in understanding these two to flashback in the next two chapters to his ideas.

The context de Certeau provides for our examination of leisure and consumption is mostly phenomenological and psychoanalytic. It is phenomenological in his focus on experience and everyday interpretation of products. It is psychoanalytic in that de Certeau saw much of the tactical work of people as unconsciously carried out. There is no doubt that he also read elsewhere in the social science of the day, for his conceptualization of strategy, for example, jibes well with many of the tenets of conflict theory.

Pierre Bourdieu

The idea of habitus, which is most commonly associated with the work of Pierre Bourdieu (1930–2002), was not, however, new to him; its prototype can be found in, among others, the works of Marcel Mauss. Mauss wrote about 'body techniques' (the physical constitution of the body) a concept that Bourdieu extended to include a person's beliefs and dispositions. For the latter habitus served as his way of resolving the difficult antinomy of objectivism and subjectivism, which has haunted the social sciences since their inception.

For Bourdieu the habitus is constituted of a set of dispositions, or schemes of perception, thought, and action that individuals acquire through their social experience and socialization (Bourdieu, 1977). Through socialization over many years each person slowly learns a set of ways of thinking, feeling, and acting which becomes firmly rooted in personality. Bourdieu argued that these dispositions – the habitus – serve as the basis for future practices of the person whose habitus it is. Furthermore, the individual develops these dispositions in response to various determining structures, including social class, family, and education. In return the individual through his habitus 'reproduces' societal structure; that is, the individual is a source of social change.

But the dispositions also form in response to the external conditions people encounter. Bourdieu called these conditions 'fields.' Society is defined by a complex overlay of distinctive fields, economic, cultural, artistic, sport, religious, and so on. Each field is organized according to its own logic, determined by the particular nature of its issues, assets, and resources. Individuals can mobilize this logic to their advantage. Therefore habitus is neither wholly voluntary nor wholly involuntary.

The habitus is Bourdieu's mechanism for moving between the determinism of society's structures – its institutions, organizations, cultural practices – and the agency of individuals who manage from time to time to operate outside those structures. That is, though the individual is socialized according to the structures, personal interests require him to navigate certain fields in life using their existing assets and resources. This experience develops the practical skills and dispositions needed to navigate effectively within and across different fields (e.g., within art as a creative pursuit and as a livelihood) and guides the choices of the individual without ever being strictly reducible to prescribed, formal rules. As a result the habitus is constantly remade by the various navigations and choices and their successes or failures.

With the postulation of neither complete determination by social factors nor individual autonomy, Bourdieu's habitus can mediate between 'objective' structures of social relations and the individual 'subjective' behavior of actors. This is possible because objective social structures are inculcated into the subjective, mental experience of agents.

Nonetheless, in Bourdieu's theory, agency is not directly observable in people's practices or in their habitus; it is directly observable only in the experience of subjectivity.

Bourdieu also described how people in power define such 'aesthetic' ideas as 'taste' (Bourdieu, 1979, p. 41). His research showed how social class can determine the likes and interests of its members, and how taste- and class-based 'distinctions' can become ingrained in daily life. The resonance of these ideas with those of Veblen is most evident, and they have the same implications for leisure and consumption considered in Chapter 2.

Some critics argue that Bourdieu's theoretic position, contrary to his claim, in fact retains the determinist flavor of structuralism. Furthermore, some of them hold, Bourdieu's habitus governs so much of a person's social make-up that it significantly limits his human agency. In Bourdieu's references to habitus it sometimes seems as if so much of an individual's disposition is predetermined by the social habitus that its component dispositions can never be altered or abandoned (e.g., Collins, 1981, p. 179). On the other hand, defenders of the Bourdieuian thesis argue that these critics misunderstand and exaggerate the conservative impact of habitus there. Bourdieu allows agency its location within the bounded structures of society and self.

As with many of the other thinkers covered in this chapter, Bourdieu relates to leisure and consumption only in a most general way. Obviously a person's habitus will include tendencies toward certain leisure activities and the consumptive purchases related to them. They will often be consistent with that person's social class, and will thereby serve as means of distinction. How habitus plays out across the vast field of leisure activities and the consumptive patterns that go with them was not a matter that concerned Bourdieu to any great extent. Nonetheless, his emphasis on human agency squares well with the fundamental assumptions of the serious leisure perspective. It is in leisure where we must look to find the richest display of human inventiveness and evidence that *homo otiosus* is far more than a mere automaton animated by deterministic, structuralist machinery.

Bourdieu brings a combined sociological and anthropological perspective to the study of leisure and consumption, for at times, he has worked from a methodological stance that is decidedly of the second discipline. Early in his career he did ethnographic fieldwork in Algeria, where he observed first-hand how people's beliefs and dispositions influence their behavior as they enact it within community social structure. Still Bourdieu read widely in the sociological theory developed before his time, including that of Marx, Weber, and Durkheim. The habitus most closely follows on Weber's *Verstehen*, an idea that emphasizes agency while clashing with the functionalist/organicist assumptions of Durkheim.

Zygmunt Bauman

Bauman, a Polish emigré who became a Professor of Sociology at the University of Leeds following an anti-Semitic purge in his native land in 1968, is well known for, among many other contributions, his original observations on consumption. He argued (Bauman, 2003, original text published in 1983) that most workers in the industrialization period of Western society were forced by circumstances to work in factories, where their unskilled labor was heavily controlled. Not so, however, with skilled workers, who were indispensable but whose skills gave them a significant measure of autonomy typical of the craft workshop. 'In a great number of cases the craftsmen were drawn into the productive process in the role of subcontractors, controlled more by the rules of market exchange than by any direct supervision of their work' (Bauman, 2003, p. 58). The owners of the factories had, therefore, to find other ways to control this small, but important group.

Consequently, compliance of these workers had to be bargained for. For one, they were paid more for their labor than their unskilled colleagues. This condition is made possible by the process Bauman dubbed 'economization of power conflict.'

> From the perspective of its [economization of power conflict] ultimate consequences, it may be depicted as the trade-off between the acceptance of the stable asymmetry of power and heteronomy inside the productive activity, and the rendering of the share in surplus open to contest. Money becomes a makeshift power substituted for the one surrendered in the sphere of production; while the experience of unfreedom generated by the conditions in the workplace is re-projected upon the universe of commodities. Correspondingly, the search for freedom is reinterpreted as the effort to satisfy consumer needs.
>
> (Bauman, 2003, p. 58)

Consumption of goods produced in the factories was a main route by which owners could establish a degree of control over these workers.

Bauman saw the unsatisfied need for autonomy among these skilled workers as placing constant pressure on the urge to consume. The result: higher levels of consumption than observed for their unskilled counterparts.

Bauman, born in 1925 and still living, is a contemporary of Baudrillard, de Certeau, and Bourdieu. While his thesis about the economization of power conflict is unique to him, he shares with these thinkers their critical understanding of consumption. Nevertheless Bauman does not link consumption to leisure, as the others do (in general terms), though the urge to consume about which he wrote could easily be interpreted to mean an urge to consume, among other things, leisure goods and services. The context he provides for the study of consumption is macro-sociological.

George Ritzer

George Ritzer, a contemporary American sociological theorist and observer of modern social life, is known for, in addition to a range of other scholarly contributions, his thesis on 'McDonaldization' (Ritzer, 1993). Its principal tenet is that many aspects of modern society have taken on the characteristics of the fast-food restaurant, as seen in the reign of four key qualities: efficiency, calculability, predictability, and control. This, he holds, is the new rationalization of social life, and a notable portion of leisure today may be described in just these terms (e.g., eating in fast-food restaurants, patronizing amusement parks, buying tickets online to a rock concert). McDonaldization, then, is a process, the spread of rationalization across modern industrial societies everywhere. The archetype of this phenomenon, Ritzer holds, is the McDonalds restaurant chain.

Though Weber's ideal-type of rationality is Ritzer's theoretic starting point for this thesis, the latter contains a distinctively modern twist missing in the former. Ritzer (1993, pp. 9–12) explains how four 'basic and alluring dimensions' of McDonaldization have propagated rationality in the food industry and elsewhere. One, McDonalds is a remarkably efficient restaurant business, as it moves from point to point in the provision of food. Two, food and service are both quantified and calculated (e.g., 'Big Mac,' 'large fries'). Three, McDonalds' products are predictable, in that a given item on the menu will be the same throughout the chain. Four, the first three are achieved through exceptional control, 'especially through the *substitution of nonhuman for human technology*' (p. 10, italics in the original).

McDonaldization subverts craftwork by stressing standardized products purveyed in measurable packages. The McDonalds hamburger and its accompaniments (e.g., fries, milk shake) offer a familiar case in point. Twentieth-century rationalization, spurred by high rates of production and the low prices they make possible, can be immensely profitable. The success of the McDonalds organization stands as proof of this. Such a scenario brings us to the question of the social and personal context of McDonaldization. Weber underscored the importance of rationalism in the rise of capitalism, and Ritzer's thesis, with its recognition of the importance of productivity and profit, attests the health of this institution in modern society. His observation, reminiscent of Marx, about the subversion of craftwork at the hands of rationalized industrial advancement further contextualizes McDonaldization.

But there are other more essentially present-day elements in the background of this process. Thus Ritzer (1993, pp. 12–13) notes that an 'irrationality' haunts the rationality of McDonaldization, in for example, the way its products contribute to environmental pollution (e.g., styrofoam packaging) and to the ill-health of regular customers (e.g., excessive salt and unhealthy cholesterol in many of the menu items sold). In addition, Ritzer observes, patronizing McDonalds has a 'dehumanizing' quality about it, as people queue up to order at the counter or drive-through wicket. 'They feel as if they are dining on an assembly line' (p. 12), as workers produce the requested order on an assembly line before their very eyes.

McDonaldization may be observed in many parts of the modern leisure industry. Ritzer (1993, p. 48) provides evidence of its efficiency from the health club business, where it is achieved by assembling an impressive variety of highly specialized exercise machines under one vast roof. He cites the package tour (Ritzer, 1993, p. 68) as an example of McDonalds style calculability, manifested in the number of sites visited (more is better) rather than the quality of each visit (which takes time). For an instance of predictability in rationalized leisure, Ritzer (1993, p. 89) turns to the modern proclivity in Hollywood for producing sequels of highly successful films. These include *Star Wars, Raiders of the Lost Ark, The Godfather*, and *Back to the Future*, each spinning off for their viewers the same characters, actors, and basic plot. The control dimension of McDonaldization of leisure may be observed, among other places, in some of today's amusement parks, where 'humans [rather than robots] do the performing, but their songs, dances, and speeches are rigidly programmed' (Ritzer, 1993, p. 111).

McDonaldization, Ritzer is convinced, is with us 'for the foreseeable future.... We face a future of ... an iron cage of McDonaldization'(Ritzer, 1993, p. 147). This situation should be resisted through personal rejection of McDonaldized businesses and organizations and by patronizing 'reasonable alternatives' (Ritzer discusses Ben and Jerry's ice cream company and the bed and breakfast industry as exemplars). A review of criticisms of the McDonaldization thesis by David Harris (2005, pp. 183–187) reveals an array of academic disagreement with it. Some analysts are uncertain of just how ironclad the cage created by this process; we must wait and see. Others hold that there is more individuality to the McDonalds restaurants across the world than Ritzer allows (e.g., Harris found alcohol available in those in France). And how many people are already resisting McDonaldization where they possibly can (before reading of or hearing about Ritzer's exhortation), in particular, by avoiding the restaurants that started it all? The McDonaldizing trend in various sectors of modern society does not necessarily mean that all members will become automatons, as the model predicts. Indeed, the later chapters of this book will show that leisure-related consumption cannot by anyway be understood in McDonaldized terms alone.

Disneyization, Etcetera

Albeit with some overlap, it is possible to discern two bodies of literature – both almost exclusively critical – that bear on the work and legacy of Walt Disney. One of them, which analyzes Disney as a man, the nature of the empire he built, and its effect on consumers (e.g., Rojek, 1993), will not be discussed in this book. Rather, mindful of limitations of space, our focus will remain broad; it will continue to center on the Western world and the forces of consumption and leisure at work within it. This focus is in line with the second body of literature, which goes by several names coined by as many writers, for example, Disneyitis, Disneyfication, McDisneyization, and Disneyization. Their writings contain analyses of the effects on society of Disney's ideas, many of those ideas finding expression in his theme parks.

Brendan Gill (1991) appears to have been the first to use the related terms of 'Disneyitis' and 'Disneyized.' They appeared in an article describing American architecture in the early 1990s, which he said reflects what is happening generally to American culture. Later Sharon Zukin (1996) briefly applied 'Disneyization' to all urban culture. In essence the idea is that more and more American cities are beginning to look like theme parks, with buildings constructed to resemble a bygone era (e.g., the Wild West, the Old South), another culture (e.g., European streetscape, Chinese architecture), or a storybook motif (e.g., Wizard of Oz, Winnie the Pooh), to mention a few. These are, first and foremost, business strategies; the aim is to make money, often by selling casual leisure experiences.

Working by analogy to Ritzer's treatise on McDonaldization, Alan Bryman gave the term Disneyization some theoretic substance. He did so by extending the idea of Disneyitis beyond urban architecture, to argue that aspects of our society are increasingly exhibiting features associated with Disney theme parks. Bryman defines Disneyization as 'simply the process by which the principles of the Disney theme parks are coming to dominate more and more sectors of American society as well as the rest of the world' (Bryman, 2004, p. 1). He holds that 'Disneyization' is characterized by four aspects, or dimensions, which may be observed not only in Disney's theme parks but also with growing frequency in many other spheres of modern Western life. They are as follows:

> *Theming* – clothing institutions or objects in a narrative that is largely unrelated to the institution or object to which it is applied, such as a casino or restaurant with a Wild West narrative;
>
> *Hybrid consumption* – a general trend whereby the forms of consumption associated with different institutional spheres become interlocked with each other and increasingly difficult to distinguish;
>
> *Merchandising* – the promotion and sale of goods in the form of or being copyright images and/or logos, including such products made under license;
>
> *Performative labour* – the growing tendency for frontline service work to be viewed as a performance, especially one in which the deliberate display of a certain mood is seen as part of the labour involved in service work.
>
> (Bryman, 2004, p. 2)

Two of Bryman's dimensions need further explanation. Hybrid consumption refers to purchasing a good or service that involves two or more different ways (Bryman calls them 'forms') of consuming it, a process that blurs the conventional distinction between the ways. Shopping, visiting a theme park, eating in a restaurant, staying at a hotel, patronizing a museum, going to a cinema, gambling at a casino, and so on exemplify various ways of consuming. Hybrid consumption occurs, for example, when people go to a museum, visit its gift shop, and enjoy a meal in its restaurant, all accomplished within the same establishment during the same session of leisure.

Performative labor is so named because it resembles a theatrical performance. That is workers serving the public are increasingly also performers, a main component being what Bryman has labeled 'emotional work.' Such work refers to 'employment situations in which workers as part of their work roles need to convey emotions and preferably to appear as though these emotions are deeply held' (Bryman, 2004, p. 104). In Disneyization the emotions are generally positive. Workers are supposed to smile, look patrons in the eye, and in this manner, convey genuine attachment to their job and the people they serve. It is hoped that the client-observers of such histrionics will then feel good about the workers and the organization in which they are employed.

Bryman says that Disneyization parallels McDonaldization, even while the two processes are different. The goal of the first is to increase the appeal of certain goods and services and the settings in which they are offered. These settings are increasingly homogeneous, which is however, one of the consequences of the second. Disneyization is thus more directly centered on consumption than McDonaldization, since it revolves around how to run a business (remember the four dimensions of the latter: efficiency, calculability, predictability, and controllability).

Bryman (2004, pp. 5–10) also distinguishes Disneyization from the related idea of Disneyfication. Richard Schickel (1986) was one of the first to use the latter term. He said that we 'Disneyfy' when we transform an object into something superficial, often done by simplifying or trivializing it. The intention is to render objects understandable to the lowest common denominator of patron, a formula Disney routinely followed with historical events, fairy tales, traditional stories, and the like. In this adulterated state it was believed that the material would be optimally entertaining, or in the language of this book, it would generate a casual leisure experience for largest possible segment of the target population.

An anthropology of consumption

Disneyization and allied processes provide a cultural context for the study of leisure and consumption, something anthropology also does, albeit in a markedly different way. Some of the previous writers covered in this chapter have flirted with the anthropological context of consumption, most notably Bourdieu, but the richest theoretic treatment of general consumption from an anthropological perspective has been mounted by Mary Douglas and Baron Isherwood (1979). The broad sweep of the study of consumption is evident in their definition of the process: 'a use of material possessions that is beyond commerce and free within the law' (p. 37). Many tribes have no commerce, but nevertheless many of their members consume material goods.

These material possessions are mostly practical; they include houses, gardens, barns, and implements. In one way or another they provide food, shelter, and protection from enemies and the elements. As importantly, these possessions enable people to make and maintain social relationships. Such goods have a dual role in providing sustenance and establishing interpersonal ties. 'This is a long-tried and fruitful approach to the material side of existence, which yields a much richer idea of social meanings than mere individual competitiveness' (Douglas and Isherwood, 1979, p. 39). Though the authors mention no names, this observation seems directed at the position taken by, among others, Veblen, Baudrillard, and Bourdieu.

What types of relationships get established and maintained through consumer goods? One type is familial. For example, a hunter who uses a bow and arrow to kill an animal for meat to feed his family is simultaneously maintaining ties with his spouse and children and possibly members of an extended group of kin. Another type is economic. Thus, a group of women working with needles and thread, using the skin of the slain animal, make clothing to be worn by members of the tribe. Links are thereby created and maintained between the women and the beneficiaries of the garments. The women also form their own relationships as partners in the process of making clothing.

Douglas and Isherwood (1979, p. 44) did not deny, however, the importance of 'private enjoyment.' It does occur, although we must recognize that such enjoyment (read consumption, some of which is not especially pleasurable) is commonly socially structured and socially interpreted. Consider a modern urban example. Some people eat alone, but do so at standard times of the day while consuming conventional breakfasts, lunches, or dinners. Others live alone in an apartment, the address of which nonetheless speaks loudly to friends and acquaintances about the financial capacity of the former to pay its high rent.

Either alone or together with other people, material goods constitute the center of an 'information system' which conveys their cultural meaning throughout the groups and networks to which the owners of those goods belong. The goods may be acquired in some kind of marketplace such as a North American department store, Middle-Eastern souk, or South Pacific trading ring, but as probable, they may not be. Rather they may be consumed through barter, borrowing, gift exchange, family inheritance, or a similar channel. Which of these arrangements is available depends on the economic foundation of the society in question.

In other words, goods are endowed with special value by consumers, and to the extent that it is shared by other consumers of the same good, this value becomes part of community culture. This 'information approach' constitutes a far larger part of the consumption process than that of acquiring and using a particular good. Douglas and Isherwood (1979) explained it in poetic terms:

> The stream of consumable goods leaves a sediment that builds up the structure of culture like coral islands. The sediment is the learned set of names and names of sets, operations to be performed upon names, a means of thinking. (p. 51)

In addition to enjoying the goods acquired, consumers enjoy sharing these names. That is, they are fond of talking among themselves about a particular 'named' good (e.g., a football match, new dress, television program, meal at a restaurant). Such sharing often includes expressing an appreciation of the good being discussed, carried out with reference to standards that all in the discussion understand. This is culture in action, as it were.

The authors' definition of consumption is consistent with Belk's presented early in Chapter 1. The one used in this book is, however, narrower than either, in that it revolves strictly around monetary acquisition, defined as either buying or renting with money a good or service. Bartering, borrowing, and similar practices have been excluded. Moreover, consumed services are of central importance in this book, whereas they were not considered in the Douglas and Isherwood monograph. For our purposes their book is especially significant for the cultural and cross-cultural contexts in which they studied consumption, both contexts enabling us to understand this process in a different light. And this holds even though the authors showed little direct interest in consumption and leisure. Later chapters in the present book will show how important their informational approach is in a comprehensive understanding of these two.

Consumption, work, and leisure

In the industrialized world of today, there is more time after work than ever. Still, it was true for awhile in the United States, as Juliet Schor (1991) concluded in *The Overworked American*, that many of them were so eager to make money to buy coveted items that they took second jobs or worked overtime whenever possible, striving in the meantime to economize through do-it-yourself projects. After meeting the multitude of obligations they had set for themselves, these drudges found they had scarcely any free time. Only four years later, however, Christine Howe (1995) claimed that this attitude was changing. Now more and more American workers were emphasizing 'reasoned wellness,' while backing off from their earlier greed and narcissism. She held that they were also starting to endow family time with the same degree of importance as work and other forms of obligated time. In a similar vein, Nicole Samuel (1994, p. 48) described the tendency observed by Schor as a 'temporary development' in the United States that stood out against the world-wide trend toward increased free time.

Overall, American research does support the claim that the after-work time of many people in that country has been growing both in amount and significance (Robinson and Godbey, 1997). But, oddly, this research also suggests that some of them feel more rushed than ever. Jiri Zuzanek (1996, p. 65) provides the clearest statement yet of this paradox. 'In general, US time-budget findings parallel those of Canada, and pro-vide additional evidence of the complexity of life in modern industrial societies marked, as it seems, by concurrent trends toward greater time freedom as well as "harriedness" and "time pressure." ' And this general trend persists despite the present tendency in a number of industries for managers to wring many extra hours of service from their full-time salaried and hourly rated employees. Yet the size of this group of reluc-tantly overworked employees is shrinking as more and more of their positions are lost in the nearly universal shuffle to organize as much work as possible along electronic lines.

Bernard Lefkowitz (1979) prefigured this trend – the willful and substantial reduction of life's work and nonwork obligations. His inter-views, conducted during the late 1970s, suggested that a small but growing number of Americans were expanding their leisure involve-ments by voluntarily accepting unemployment, partial employment, or early retirement. Two years later Daniel Yankelovich (1981) confirmed these impressions in a nationwide survey of Americans. Today, the evi-dence clearly supports the proposition that the work ethic of old is waning in intensity, even in North America, the home of the largest number of its adherents by far.

Now, paralleling the tendency to voluntarily reduce obligations is a much stronger force: the technologically driven, involuntary reduc-tion of paid work. In *The End of Work* Jeremy Rifkin (1995) described

the current decline in the size of the global labor pool and the traditional market economy and how both forces are now pushing ever
larger numbers of people toward greater free time at an alarming
rate, whatever the individual's interest in reducing his or her level
of work. As the 21st century dawns, a wide variety of employable
men and women are finding that their job opportunities have shrunk,
sometimes to nothing at all. Behind these unsettling trends lie the
powerful forces of what Rifkin called the 'Third Industrial Revolution':
the far-reaching effects of electronic technology as manifested in the
micro-circuitry of computers, robotics, telecommunications, and similar
devices.

The Information Age has dawned. In it, Rifkin observed, these tech-
nologies will continue well into the 21st century inexorably replacing
workers, either directly or indirectly, in virtually every sector of the
economy, including manufacturing, transportation, agriculture, government, and the retail and financial services. New jobs will be created
in significant number only in the knowledge sector, in science, computing, consulting, education, and the technical and professional services directly related to the new technology. Rifkin said this sector
will compose no more than 20 percent of the workforce. Jobs lost in
the other sectors will be gone forever, offset very little by the comparatively small number of jobs generated in the knowledge sector.
And occupational retraining is no solution, for the people in line for
such retraining generally lack the necessary educational background on
which to build the skills and information they would need to work in
the knowledge sector. In short, regardless of the orientation of the modern man and woman toward life after work, he and she now seems
destined to have much more of it than ever. Two other books published by Stanley Aronowitz and William DiFazio (1994), and Ann
Howard (1995) indicate that Rifkin was not alone in observing these
trends.

Yet these writers have failed to address themselves directly to this
question. Instead they advance two broad observations, one tantaliz-
ing, the other frightening. They predict that the Information Age will
offer greatly expanded opportunities for leisure and personal development by way of it and that this Age will offer free time far in
excess of the typical person's capacity to use it constructively.[1] In the
eyes of the person trying to adapt to a world buffeted by momentous change, these observers paint a picture of life after work in
the Information Age that is both too hazy and too unsettling for
comfort.

The economics of modern leisure

One prominent aspect of the institution of leisure is its economic base; especially in the West, people spend huge amounts of money buying or renting leisure goods and services. Linda Nazareth, who devotes a chapter to the monetary impact of leisure, writes that a 30 billion dollar craft and hobby industry thrives in the United States (Nazareth, 2007, p. 204). To this figure, we must also add the substantial sums she says are generated annually in entertainment, reading, cooking and eating out, adult education, sport and fitness, gambling, and travel. In short, leisure in Western society is a big business.

It is a big business in good part because many people now have the personal resources of time and money to spend on leisure goods and services, doing so in quantities having noticeable consequences. Moreover, says Nazareth, who analyzes the North American situation, the long-term forecast is for both of these resources to expand, spawning widespread change throughout entire societies. The later baby boomers (the boomer generation runs from 1946 to 1964) started this trend, which members of Generation X (born between 1965 and 1976) and Generation Y (born between 1977 and 1999) are feeding with ever greater enthusiasm. This trend in growth of money and time for leisure is pushed by an increasing desire sweeping through recent generations for more leisure and less work. To be precise, they seek a much more even balance of the two than heretofore seen.

Nazareth (2007) looks to the 'Gen Ys' to come closer than any of the preceding generations to the popular ideal of a decent work/leisure balance.

> Generation Y is the group that might catapult North America into the leisure economy (p. 80)... [This generation] has experienced a different reality than Generation X in the job market. By and large, they have grown up not just knowing decent economic times, but have also been told that the world has in fact been waiting for them (p. 86).... So perhaps it should not be a surprise that newly minted Generation Y graduates tend to be pretty picky about what they want in a job.... And there it is in a nutshell: Generation Y treasures their leisure time. (p. 88)

One of the results of this shift in attitude is a weakened effectiveness among managers in such fields as law and medicine for wringing long hours of employment from employees of this age. Nazareth also notes that this generation is endowed with a strong volunteer ethic, partially explained by mandated volunteer programs they participated in during their school years.

In fact, some of the younger Gen Xs and Gen Ys dream of completely replacing work with leisure. But working or not the leisure contemplated has to be 'meaningful,' with a good portion of it involving their children (Nazareth, 2007, p. 95). Such an orientation is found in both men and women, and is not unique to these two generations, even if it is most evident in the second. The appeal of 'slacking,' as Nazareth calls the preference of leisure over work, has also been observed in some baby boomers, who when forced to leave the labor force, have developed a love for leisure (p. 98).

This, then, is the economic context in which the field of leisure and consumption is increasingly embedded. The perspective on the leisure economy brings to the fore the place of work, its meaning, and its importance *vis-à-vis* free time and the activities pursued there. Nazareth (2007) traces many of the diverse consequences this shift in interest has for the larger society, some of which we have just discussed. In a nutshell, we in the West are evolving toward a 'leisure economy,' she says, or an economy dominated by a search for an attractive balance of work and leisure. Be that as it may, Tim Robinson (2008) believes that Western societies remain locked in 'a vicious cycle of overwork and over-consumption.' Until this pattern is broken the leisure economy will fail to develop fully.

Meanwhile in the present serious, global, economic decline, the Gen Ys are now said to have changed from demanding 'What can you do for me?' to announcing 'Here's what I can do for you' (Jessica Buchsbaum quoted in *The Economist*, 2009, p. 47). Jobs are scarce today, so for many of these workers, slacking is no longer an option. And, in this period of financial gloom, what is the fate of the propensity to overconsume? People are presently consuming much less than they did in the economically good times of the recent past. Will this new trend continue once economic health returns, which for many economies, as of this writing, is predicted to be several years off? Or will the vicious cycle of overwork and overconsumption be broken, possibly forever?

Conclusions

Over the years the scholarly world has created a rich and variegated set of contexts within which to explore leisure and consumption. Modern thought on the question began with the historical and socioeconomic contexts Marx used to examine society. At the same time and later Simmel, Lefevbre, Baudrillard, Bourdieu, Bauman, and Ritzer framed consumption in, for the most part, macro-sociological terms. Simmel was arguably the most social psychological of these thinkers. Baudrillard's approach to understanding consumption also contained cultural and philosophical elements, while de Certeau looked on the process through phenomenological and psychoanalytic lenses. Finally Bourdieu and Douglas and Isherwood brought an anthropological slant to the study of consumption.

Consumption for some of this group of thinkers was treated of largely, if not wholly, without direct reference to leisure (e.g., Marx, Simmel, Baudrillard, de Certeau, Bauman, Douglas and Isherwood). For others leisure does fall within their purview, albeit only very generally (e.g., Lefevbre, Bourdieu, Ritzer, Bryman). Interestingly, some of the writers in this second group do differentiate certain kinds of work, most often those of skilled and unskilled labor, while failing to analyze leisure as carefully. Instead they view the latter as either a homogeneous block of activities or a sort of institution.

Many of the contextual ideas put forth by the writers covered in this chapter are consonant with those of several of their colleagues discussed in Chapter 1, critical of the consumptive order they analyzed. But, in tune with the positiveness of leisure, some of the first, without directly saying as much, do find some positiveness in consumption. This is evident in, for example, Lefevbre writings where he stated that it is in everyday life where the self is formed and the creative potential of each person is realized. Baudrillard saw much the same process at work in the consumer culture, while de Certeau viewed it operating through consumer tactics. All three men were, in effect, underscoring the importance of the role of personal agency in everyday consumptive life. On another plane Douglas and Isherwood's informational approach highlighted the joy of discussing consumer products and sharing thoughts about them. Nazareth's predictions about the coming of the leisure economy also have a positive ring. In short, it would be inaccurate to say that, since Marx, thought on consumption has been wholly critical, wholly negative, even if this strident motif of that body of ideas is what most outsiders to this area seem to see and feel first.

We may only speculate about the reasons for the incomplete conceptualization of leisure shown by the writers presented in this chapter. For, alas, there has been no discussion of these reasons of which I am aware. I wish to suggest, then, that one reason why leisure was given short shrift by the thinkers who were born and raised and had obtained their doctorate before World War II was that there was relatively little of it compared with the decades following this historic turning point. This observation should be nuanced, however, with the fact that Europe, Canada, Australia, and New Zealand, among other Western societies, lagged the United States in their transition to becoming ones characterized by significant free time. For these observers leisure was, in effect, out of sight and hence out of mind.

Second, the dominance of work in the West both as activity and as institution – even though this dominance varies in its strength across the component countries – may also have blinkered the way a number of the men and women covered in this chapter perceived leisure. Even in the 21st century many a social scientist (outside the field of leisure studies) still sees it in simplistic, universalistic, common-sense terms as out-and-out hedonism, or somewhat more analytically, as mere casual leisure. In many intellectual circles, it appears, both hedonism and casual leisure have bad names, an attitude rooted in the conviction that they are a waste of time, source of evil, blight on the development of the human personality, and similar deprecations. Certainly not subjects on which a thinking man should devote his scholarly energy, some would argue.

If they accomplish little else the next two chapters should show just how erroneous and detrimental this stereotype is.

4
Phase One: Shopping

Phase One centers on the activities and conditions actually or possibly leading to acquisition of a good or service. The underlying model here states that the monetary act of acquiring (i.e., purchasing or renting) a good or service stands between Phases One and Two. Phase Two refers to the activities following on the acquisition of a good or service engaged in by the person making the acquisition. What demarcates the two phases is actual payment and thereby possession of the purchased good or service.[1] The field of consumption studies has focused primarily on Phase One as well as on a matter extraneous to the model: the effects of consumption on individual and society (see Chapters 2 and 3). What people do with their purchases, a matter of discussion for Phase Two of the model and Chapter 5 of this book, has been of much less concern in this field.

In the present chapter I distinguish between *obligatory consumption* and *leisure-related consumption*, even while shopping, as leisure occupies center stage throughout. Obligatory shopping and consumption serve thus as a comparative background. This said shopping may occur as work, leisure, or nonwork obligation, or as any combination of these. For most people in Western society going shopping is, some of the time, done to meet necessity. The family needs food on the table, the bread-winner needs a new suit, or the children need supplies for school, all examples of obligatory shopping both inside and outside the sphere of work.

The following line from one of Arthur Miller's plays says is a lot about the other, leisure, orientation toward shopping:

Years ago a person, he was unhappy, didn't know what to do with himself – he'd go to church, start a revolution – *something*. Today you're unhappy? Can't figure it out? What is the salvation? Go shopping.

(from: The Price, Act 1)

The kind of shopping considered in the preceding paragraph is not the kind that Miller had in mind, for shopping out of need is unlikely to assuage unhappiness. Indeed, it seems quite capable of causing it.

In brief, shopping varies noticeably depending on whether the shopper sees it as something done for work, leisure, or to meet a nonwork obligation. By and large, discussion will be confined to leisure and nonwork obligation, the shopping at work being much the same as these two except that it is part of a person's job. For instance, an employee might feel it is a nuisance to have to buy coffee periodically for daily brewing in the office pot, but find pleasure in the assignment to buy flowers for the annual Christmas party.

The nature of shopping

Allowing for the exceptions considered shortly, the activities of Phase One may be treated of here under the rubric of 'shopping.' Shopping is what people do when they 'shop,' which is to go to a shop, store, or office to view or purchase a good or service, if not both. Broadly conceived of, today's shop is increasingly an online entity, even if the traditional shop seems destined to continue to play an important role in consumption. And people still contact shops at times, using the telephone and ordinary mail to discuss goods and services or make purchases. This is how many of us buy insurance policies, arrange for land line telephone service, reserve tickets to the opera, order a pizza for home delivery, and the like.

Shopping as just described is a manifestation of human agency, of personal initiative to go to shops and their equivalent to look at and possibly buy something. The decision to go shopping may be forced by circumstances (e.g., the family needs food, the shopper needs a coat), but the shopper decides within such constraints when, where, how, and perhaps, with whom to buy the needed good or service. So we may say that shopping is characterized by choice, albeit one typically hedged in by many limitations. There may be only a few items or services to choose from, there are often upper limits to what a buyer can or will afford, there may be constraints forced by geographical availability of the 'shop,' there may be ethical restrictions on purchases (e.g., green considerations, despised manufacturer, religious restrictions), to list but a few.

With human agency and restricted choice as ineluctable conditions of shopping, it is important to note that this process, widespread as it is in many parts of the world, fails to describe fully the acquisitive side, or Phase One, of consumption. That is, the modern consumer occasionally purchases goods and services beyond the sphere of the shop (in the widest sense), where agency is impossible and no realistic choice of the good or service exists. For example, in many parts of the world, people may buy natural gas and electricity for their homes, but only from a public utility. Similar monopolistic arrangements exist often for purchasing public transit tickets, joining a trade union in a union shop, paying into a mandatory company health plan, contributing to a mandatory governmental retirement program, and the like. Such is not shopping, even while it is consumption.

These instances of consumption outside the domain of shopping – consumptive acts that are nevertheless part of Phase One – are not covered in this chapter. This is not to argue that they are unimportant. Much to the contrary. Monopolistic control of certain goods or services is infamous for presenting a variety of problems (e.g., inflated prices, inferior service, poor-quality goods) and people spend large amounts of money on many of them. It is rather that such consumption falls well beyond the realm of leisure and the scope of this book. Meanwhile shopping, given its agency and limited choice, harmonizes nicely with our definition of leisure (see Chapter 1 for a discussion of agency, choice, and leisure), even if by no means all shopping can be conceived of as leisure activity.

A short history of shopping

Historical information about shopping is, for the most part, limited to this practice as found in the West.[2] Here shopping has always been done mainly in cities and towns, to the extent that people living in such places are unable or unwilling to grow or raise their own food, make their own clothing and shelter, and otherwise directly provide for the necessities of life. So it should come as no surprise that, for a long time, some people shopped – albeit mostly the elite, it seems – patronizing urban street-side stalls where artisans sold goods that met certain of the needs and wants of the first. Such was evident even in ancient Greece and Rome. From the slim evidence there is on the matter, the *demos* of ancient Greece, who were independent craft workers, operated in much the same way as some of the small business crafts people do today. Their clients came from the top stratum.

Clarke, Doel, and Housiaux (2003, p. 17) argue that, whereas we can sometimes identify consumptive patterns back through history, it is best not to view any particular period as an especially important turning point in consumptive practices.[3] Rather the consumer society we know today springs from numerous 'intersecting and overlapping changes.' Thus, after ancient Greece and Rome, there is evidence of consumer activity around 1300. Other evidence gathered between approximately 1540 and 1730 in England shows that consumption during that period grew ever more varied, in that more and more people from increasingly different backgrounds began to enter the marketplace. Late in this span of time, consumption in England came to be guided by *sumptuary law*, which stipulated which classes of people could buy which goods and services. But, subsequently, as the modern period progressed, this practice began to break down. Still the individualist nature of today's consumption evolved only slowly during this era of history.

Across much of time only a small proportion of the population consumed in a manner familiar to us in the present. Usually this was the elite. Meanwhile the rest of society subsisted – made and grew their own necessities – leaving little or nothing to exchange for goods or services they might consider less than essential. According to standards of the day, these less-essential goods and services were 'luxuries' (we will discuss later in this chapter the difference between wants and needs). In general, the history of consumption is one of more and more people becoming able to buy more and more nonessential goods and services, with the greatest surge in this trend occurring after World War II.

America, a bastion of world capitalism, was in the vanguard of transforming consumption into a mass practice. Herbert Ershkowitz writes that:

in colonial America [between 1600 and 1776] in each of the seaboard cities, master artisans created and sold jewelry, furniture, silverware, and other items from the first floor of their homes. Streets of the cities contained a constant parade of people buying and selling from these artisan shops or from street vendors. Shopping was a major part of life in the cities. At times, the sheer pleasure of buying became addictive. In correspondence, husbands and wives complained that their mates' buying habits were causing financial problems. Benjamin Franklin,

for example, criticized his wife Debra, for spending too much money
on clothes and furnishings they did not need.

(Ershkowitz, 2004, p. 256)

The leisure side of much of this activity is evident, but only to the
extent that it failed to become addictive. For in addiction *homo otiosus*
can no longer control his addictive behavior; it is now coerced by
irresistible habit.

Ershkowitz (2004, pp. 256–259) goes on to describe the fashionable
shopping streets in the early 19th-century American cities, areas frequented by the elite and upper middle classes. He labels as America's
'Golden Age of Shopping' the period running from the 1870s (the Civil
War ended in 1865) to just after World War II, during which large
department stores were built in every major city. He also describes
in detail the leisure found in these places, where people would meet
friends for a day of shopping (including window shopping), admire
displays, and take in special events such as art exhibits and orchestral
concerts, visit with Santa Claus, and attend style shows. Meanwhile
in much of 19th-century Europe, entrepreneurs constructed covered
arcades filled with shops. Like the American department store they
attracted an increasingly growing and varied clientele (except during
the Great Depression).

Both types of emporia were precursors to the modern shopping
mall. For example, following World War II in the United States the
large center-city department store began to decline in importance as
a response to, among other forces, the expansion of suburban living
and the arrival nearby of large shopping malls. The department stores
quickly established branches in these places, but many of them have still
found it difficult to maintain the commercial vigor they once enjoyed
in their main stores 'downtown.'

Paralleling these developments has been the emergence of specialized
shopping opportunities in some of the urban ethnic enclaves enrich-
ing many of the world's larger cities, most notably, their Chinatowns
as well as their Greek and Italian communities. In them outsiders
mingle with insiders to buy special foods, articles of clothing, traditional decorations, ethnic meals, and the like. These experiences are
enhanced through personal interaction, however brief, with members
of the other culture. Additionally, with the growth in the past 50–60
years of national and international tourism, shopping, including that
done in these urban enclaves, has become a major component of this
genre of leisure activity.

Types of shopping

Before embarking on a discussion of types of shopping, we must introduce some guiding conceptual distinctions. First, note that shopping is viewed in this book as an activity. I defined *activity* earlier as a type of pursuit, wherein participants in it mentally or physically (often both) think or do something, motivated by the hope of achieving a desired end. Life is filled with activities, both pleasant and unpleasant: sleeping, mowing the lawn, taking the train to work, having a tooth filled, eating lunch, playing tennis matches, running a meeting, and on and on. Activities, as this list illustrates, may be categorized as work, leisure, or nonwork obligation. That is, at the activity level, all of everyday life may be conceptualized as being experienced in one of the three domains: work, leisure, and nonwork obligation. It follows that all of everyday consumer life may also be conceived of as such. One might ask at this point if our existence is not more complicated than this. Indeed it is, for each of the three is itself enormously complex, and there is furthermore some overlap in the domains.

Gregory Stone (1954) is quite possibly the first scholar to develop a typology of shoppers, or more precisely, a typology of their orientations toward shopping. This scheme emerged from research he conducted for his master's degree at the University of Chicago. Stone interviewed over 100 women in that city, asking them about their attitudes toward shopping and focusing in particular on their reasons for choosing one kind of retail outlet over another. The results of the study enabled him to identify four basic orientations toward shopping. They were evident in the 'economic' shopper, for whom the primary considerations were price and quality, usually as found in large chain stores, and the 'personalizing' shopper, who regarded such economic criteria as of secondary importance compared with the opportunity for social interaction. This type was especially keen on establishing quasi-primary relations with the personnel of local independent retail institutions. The 'ethical' shopper claimed to apply moral principles when choosing retail outlets. In particular, she preferred small businesses compared with larger operations, which she felt threatened the housewife's way of life. The 'apathetic' shopper conducted her activities out of necessity; she sought to minimize her time in the marketplace. Most of the women in this latter type were elderly.

The obligatory consumer activity discussed above, the kind of shopping done by Stone's apathetic type, and possibly that of the economic type, are evident in what Robert Prus and Lorne Dawson (1991) called 'shopping as work.' For the sake of consistency with the activity framework just set out, I will abandon this locution in favor of the expression *shopping as obligation*. For obligations that are in some way disagreeable also abound outside the sphere of work as a livelihood, and some of them can force people into the marketplace in an effort to meet them (e.g., shopping for drugs, food, petrol). Shopping as obligation was described by Prus and Dawson's Canadian interviewees as 'laborious,' a sentiment that arises from having to make purchasing decisions the latter found difficult, frustrating, monotonous, or unavoidable.

> Conversely, shopping tends to be viewed as a more laborious activity when one experiences a confused, constrained, and irrelevant sense of self in shopping situations. When people face undesired ambiguity and frustration, there is a sense of the *incompetence of self*. When they experience closure and disenchantment, *the loss of self-direction* becomes more pronounced. And when people encounter boredom, there is an emptiness or *meaninglessness of self*.
>
> (Prus and Dawson, 1991, p. 160)

Rachel Bowlby (1997, p. 102) points out that obligatory shopping (she calls it 'doing the shopping') is routine, involves mostly shopping for food, and is generally regarded as a chore.

Prus and Dawson and Bowlby contrasted shopping as obligation with 'shopping as recreation' or 'going shopping.' In keeping with the activity framework I will substitute these terms with the broader phrase of *shopping as leisure*. For Prus and Dawson (1991, p. 149) such commercial activity is distinguished by 'the overall sense of enjoyment one obtains while shopping,' where a positive sense of self is always present. Stone's personalizing shopper fits this description, as does the ethical type in the sense that, compared with large businesses, the small ones lend themselves to deeper, pleasant interpersonal relations. This is also shopping as Arthur Miller saw it – the contemporary mode of salvation.

This two-fold typology of shopping as obligation or leisure (and hence of shoppers as motivated thus) has been empirically validated in several studies. Thus, seemingly working in parallel with Prus and Dawson (1991), Pasi Falk and Colin Campbell (1997) have covered the same ground, setting out the same two-fold typology and underscoring for both types their impact on self. They also list in their bibliographic appendix a number of their earlier papers bearing on this same conceptualization. Most recently John Robinson and Steven Martin (2008) have analyzed 34 years of data collected by the General Social Survey in the United States in an effort to identity the activities in everyday life that make people happy and those that make them less so. Among their findings that bear on this chapter and that are clearly related to consumption are those showing grocery shopping to be considerably less enjoyable than 'other shopping' and eating meals away from home. The first would seem to be disagreeable obligation, while the second two are leisure activities.

Knowledge and shopping

Shopping to fill an obligation, however unpleasant, proceeds at times from a basis of considerable knowledge about the product sought and the commercial circumstances in which it is sold. Of course, this is usually not true in any fundamental sense of buying a chocolate bar or a tank of gasoline, for example. But it is definitely true of sophisticated (as opposed to naïve) buying of, say, a house, automobile, home sound system, or set of banking services. Consumers' organizations that analyze popular products and services and publish their findings so the public may be better informed exemplify the high level of product and market knowledge that shoppers need if they are to make the most informed purchases possible. In short, shopping done with utmost effectiveness often requires substantial of knowledge of the product and its market. Such shopping may be conceived of by the shopper as either obligation or leisure. No matter which it may well be a significant source of pride.

Mica Nava (1992, p. 74) believes that this acquired knowledge is a special attribute of the modern woman, who typically has more time for shopping and more reason to do it than the typical modern man. The work role of the latter often requires him to stay put outside the marketplace in such places as offices and factories. Nonetheless both men and women may be proud of their consumer knowledge, which for instance, helps them avoid long lines in the supermarket, overpriced drugs in the pharmacy, or inferior products at the hardware store. Notwithstanding such expertise, Colin Campbell's (1997) study of a sample of male and female shoppers in Leeds revealed that the women were much more likely to prefer shopping to other forms of leisure, including going to the cinema or to a restaurant. In contrast the men more often described how they 'hated' or 'disliked' shopping.

The role of knowledge in shopping is considered further in Chapter 5 in the section on facilitative consumption and project-based leisure.

Gender and shopping

Nevertheless the relationship between gender and shopping is more complicated than the difference, just examined, between the two sexes according to knowledge base and time available for frequenting the marketplace. Nava (1997) holds that not only do women play a key role in shopping they also act as agents of modernity. That is, such commercial activity is central to our understanding of modern society, many facets of which are highly gendered. Consider the modern department store. It is designed for women and heavily used by them, while also being a significant public institution in the contemporary city. It forcefully belies the claim that the activities of women are confined to private spaces, notably, the home. Female shopping, then, is no trivial matter, as some scholars have been wont to claim (Nava, 1997, pp. 58–59), but rather a main indicator of urban life in the present.

Campbell (1997) learned in his study that the women sampled were much more inclined than the men to 'browse,' or window-shop. The men were not indifferent to price, but if finding a better price required them to 'shop around,' they were more likely than the women to eschew this approach by purchasing the good or service in question at a convenient location. This they would do even if they had to pay a higher price. In other words, men were more likely than women to see buying as an obligation. In fact the men browsed, but they limited this activity primarily to technical goods such as cars, computers, and do-it-yourself equipment. Consonant with their outlook on shopping the women were more inclined to define it as leisure activity. These attitudinal differences, Campbell found, lead one sex to occasionally belittle the other for their 'curious' approach to acquiring goods and services in the marketplace and what is felt as leisure or obligation in this regard.

He also found that 'joint shopping' is not uncommon, although it is mostly limited to purchasing expensive durable items, for example, a bed, car, or house. This aspect of shopping has received little scholarly attention (Hewer and Campbell, 1997, p. 191). Questions in this area include who chooses the retail outlets to visit, which models to examine, which features of the chosen model to give priority to, and how much to pay for it. Hewer and Campbell's review of the scant literature there is suggests that women tend to dominate in buying household appliances, whereas men dominate in choosing the family car.

Rachel Bowlby (2003) examines another gender-related feature of shopping, one much in tune with a later section in this chapter on creating wants, which is how stores visually exploit the pleasures of narcissism more actively for women than for men. This Freudian approach to differentially marketing goods to the two sexes is particularly evident in shop windows. Here, she says, a powerful relationship exists between femininity and commerce. It has to do with an internalized personal image of oneself as feminine or masculine that is built into the displays and then reflected back in those displays to the (predominantly female) shopper.

Shopping as leisure

The spotlight in this chapter is on shopping as leisure, shopping that people like to do. Several observations may be made on this free-time activity. First, *window-shopping* is by far the best-known expression of such leisure. Here the shopper enters the marketplace for the enjoyment of seeing displays, looking at different creations and packaging of consumer items, fantasizing perhaps on how these might fit in that person's life, and so on, all without direct intention of buying the item of interest. This is casual leisure of the sensory stimulation type, in which curiosity plays a central role (Stebbins, 1997a). True, window-shoppers may buy something they have seen, perhaps because they have discovered a need for it, it has decorative value for home or office, or they have been looking for the item for some time and now, unexpectedly, have finally found it, even though they had not set out that day intending such.

The *flâneur* of 19th-century Paris was, among other things, an inveterate window shopper. The quintessential man of leisure, he took to the city's streets in search of casual leisure satisfaction armed with a special sensibility for observation.[4] This common figure of the 19th century has been richly described by Walter Benjamin (1973) in the 20th century by way of a critical analysis of Charles Baudelaire's poems on crowds (multitude) and solitude in Paris. The *flâneur* was at once an amateur detective, commentator on urban hustle and bustle, and explorer of daily city life. His window-shopping, which occurred as he strolled along the avenues and boulevards of Paris, was said to have distracted him. The glassed displays in the new arcades and department stores interrupted his movement through the town. The *flâneur* was not much of a buyer, however, for he would have been destitute had he tried to purchase even a little of what he saw in the shop windows. And as Gregory Shaya (2004) points out he was no *badaud*, no gawker. The *badaud* is of the crowd, the multitude, whereas the *flâneur* is a loner, someone beyond the crowd walking in solitude in the big city.

Today window-shopping is now a main part of modern urban tourism the world over. As Bill Martin and Sandra Mason (1987, p. 96) have observed: 'shopping is becoming more significant to tourism, both as an area of spending and as an incentive for traveling.' Today no tourist guidebook of a world city would be without a section on shopping, typically running from high-end clothing and gadgetry shops to low-end flea markets and bargain stalls. Moreover, small communities, even villages, when sufficiently interesting in themselves and within the orbit of a larger tourist zone, tend to be bristling with tourist-oriented shops, several of them purveying local arts and crafts. And note how window-shopping is fostered by the ubiquitous gift and souvenir boutiques that are strategically situated to attract and tempt patrons as they exit today's museums. And then there are the flamboyant shopping sections of the big international airports. They combine the enticements of duty-free bargains with the tonic of window-shopping to passengers about to endure the tedium of a lengthy flight and those who have just escaped this state after disembarking one.

As for shopping centers they, too, may be places to visit and possibly drink coffee or eat something, places sometimes frequented for reasons other than to make a purchase in a retail shop (Morris, 1993). For instance, people arrange to meet friends or relatives here or punctuate their day with some people watching or salubrious walking. Some window-shopping might still occur, even if that was not the principal motive for the visit that day.

The basic leisure motive for window-shopping begins to blur when shoppers mix purpose with curiosity. For example, tourists, in particular, may also want to buy a souvenir of the place visited or bring home a gift for someone. So while window-shopping these people are also on the watch for items that will fill one requirement or the other. In fact, if toward the end of the trip they have not yet found what they want, shopping may become more exclusively purposive, perhaps even obligatory (now a need), as its leisure character begins to fade in the face of necessity.

Serious leisure

Another type of shopping as leisure is that done in service of a serious leisure interest. Amateurs and hobbyists, in particular, must occasionally buy goods, the purchase of which can be most pleasant. A horn player sets out to find a new and better horn, a coin collector goes shopping for missing parts of his collection, a kayaker patronizes her local dealer to buy a new, lighter, and more streamlined boat. The immediate outcome is the prospect, made possible by the purchase, of better and more fulfilling execution of the hobbyist or amateur passion. Furthermore, the process of purchase itself commonly proceeds from a background of considerable knowledge and experience relative to the best products and their strengths and weaknesses. Such knowledge is central to the development of a positive sense of self, which Prus and Dawson argued can emerge from some kinds of shopping done for leisure.

Still, there are times when serious leisure enthusiasts must also engage in some obligatory shopping of their own, such as making the occasional trip to the store that sells art supplies to buy paint or brushes or to the string instrument shop to arrange for repairs to a violin. The main purpose of such shopping is likely no more pleasant than certain routine purchases are for many people (e.g., buying petrol, using banking services, having prescriptions filled), though mercifully, it occurs much less often. Moreover, the tedium of such errands may be alleviated in part by doing some window-shopping while in the store. So the painter might browse the easels or the selection of frames. The violinist might look over the violins that are for sale or consider various musical accessories like metronomes or music stands.

Distance shopping

Discussion to this point has revolved around people who do their looking and buying in stores. Yet there are two types of shopping as leisure that are essentially window-shopping, but are nonetheless conducted at some distance from the 'shop' itself. Thus, for decades in many countries, consumers have also had, in certain sectors of the economy, one or more opportunities to shop by mail-order catalogue. And for some people, especially those living in rural areas, perusing catalogues has long been a substitute for viewing real displays of merchandise, which are most extensively available only in faraway cities. Today, at least in North America, some firms still regularly publish catalogues portraying in appealing detail much of what they sell in their stores, even if only a fraction of these enterprises still offer mail-order service. Rather their marketers hope that window-shopping by catalogue will kindle a desire to buy something, leading sooner or later to a trip to a nearby outlet to appease the desire. My informal observations suggest, however, that these catalogues are valued by many people primarily for the window-shopping that the catalogues offer. Years may pass before they purchase something displayed there. In the meantime the catalogues make for interesting, casual leisure reading.

These days, however, shopping appears at least as likely to be done by computer as by catalogue using the local telephone or postal services. The World Wide Web now presents a huge range of shopping possibilities. Apart from the sites that offer vast selections of items for purchase online, are those sites that communicate information about products that would-be consumers will have to acquire at a nearby store. As on the street and in the catalogues, pure window-shopping may also occur on the Web, exemplified by browsing sites that present, for instance, what the automobile makers or computer manufacturers are selling. As elsewhere the intention here is to satisfy curiosity through entertainment or sensory stimulation; it is casual leisure.

The context of shopping

Up to now we have been considering the shopping experience, felt either as obligation or as leisure. Attention has been on the core activity of shopping in its various forms and types. But shopping takes place in a larger social and physical context, which can affect how it is defined by the shopper, as pleasant or not. In supermarkets, at least many of those in North America, part of that context is the snack bar, a relatively new addition to such places. And it would be interesting to know whether it works as intended: to reduce the sense of laboriousness, so widely felt in this shopping experience. In other words, how many shoppers take time out for coffee or a snack (or carry either with them as they shop) while nonetheless trying to put this chore behind them? While the snack bar can be defined as an attempt to inject a few moments of leisure into one of life's routine obligations, most of the social and physical context in places like supermarkets and department stores seems, for the most part, to be designed only to reduce their unpleasantness. Such measures seem to stop short of transforming into leisure this kind of shopping.

That is, less negativeness cannot logically generate a measure of positiveness, because the latter is spawned only by activities that make life worthwhile (see Chapter 1). Thus a sufficient number of check-out counters (properly staffed) helps reduce the frustration that comes with queuing up over long periods of time. Accessible merchandise with prices clearly marked can accomplish the same thing, as can a layout of aisles that facilitates easy movement for customers. The ever growing list of amenities gracing the modern shopping center can easily be understood in similar terms – as attempts to transform shopping into an enjoyable undertaking. But do they succeed? Some careful research on the matter is in order.

Arthur Miller published *The Price* in 1969. It seems that even in those days consumerist shopping was looked on as leisure-related salvation from unhappiness. Today the tendency to go shopping as leisure activity appears to be still more pronounced, even while the obligatory form still haunts nearly all of us. Of note here is the fact that *Merriam Webster's New Collegiate Dictionary* (11th edn) traces origin of the word 'shopaholic,' defined as one who is extremely or excessively fond of shopping, to 1983. The *New Shorter Oxford Dictionary* (5th edn) adds that shopaholics are compulsive shoppers. But when shopping becomes compulsive, the leisure quality of such activity vanishes, subtly but inexorably overridden by psychological anxiety and lack of self-control.

Observing that participation may be coerced brings us full circle to the problem of obligatory shopping, which it turns out upon closer examination, is not always unpleasant. Put otherwise leisure may be obligated. An obligation is an attitude or act of refraining from doing something because the actor, though not coerced, still feels bound in this regard by promise, convention, or circumstance (Stebbins, 2000, p. 152). Obligation, which may be pleasant or unpleasant, is both a state of mind, an attitude – a person feels obligated – and a form of behavior – a person must carry out a particular course of action. Personal care, child care, yard care, housework, and shopping are common nonoccupational obligations, while one's paid occupation, if employed (or self-employed), is a central work or occupational obligation. Pleasant obligations are sometimes part of leisure, as when a volunteer is obliged to serve at a certain time of the day or a football player is obliged to play in a game.[5]

And they may be part of shopping. Shopping as obligatory leisure is probably relatively rare, but yet, consider the fulfillment some people experience when buying, for example, a new house or automobile. Let us assume that they are not upset by the financial implications of such a purchase and that they have worked up a solid knowledge of the product and its market, carefully searched out the best buy, and realized the best deal, all of which can be deeply rewarding. The purchase was necessary for these shoppers, however, for their old car had been demolished in an accident or their old house sat on land over which a new highway would soon be built. By the way this leisure is most accurately classified as 'project-based leisure.' As stated in Chapter 1, such leisure is short term, moderately complicated, either one-off or occasional though infrequent, creative undertaking carried out in free time. It requires considerable planning, effort, and sometimes skill or knowledge, but for all that is neither serious leisure nor intended to develop into such.

Created wants

It is time we come to grips with a classic pair of concepts in economics that bear substantially on the contents of this chapter, notably *need* and *want*. We have been rubbing comfortably along until now with loose usage of these two ideas, which in this chapter, however, demand more careful application. In traditional economics a need is seen as having a physiological or biological basis. As such it is essential for maintaining life, for instance, needs for food, air, water, shelter, clothing, and sleep and, let me add following Robert Wilensky (1978), essential for meeting an obligation. A want, by contrast, is significantly less essential to life; it is a desire or wish, which when fulfilled makes our existence more agreeable, possibly even more worthwhile.

Yet, there are times when need and want are not so easily separated. The obligatory, nonphysiological/biological need is exemplified in such situations as: John needs to help Mary move because he promised this service; Mary needs to attend a meeting because she promised to chair it. But, in line with discussion in the previous section, these two instances of obligation might be seen by John and Mary as pleasant. In such cases they both need *and* want to fill the obligation.

Much of what has been said so far in this book has been concerned with wants and their fulfillment. Moreover, it was observed in the preceding chapter that some wants start out as needs, but for some people, get transformed by commercial interests into desires. These desires when realized make life more agreeable. This transformation further shows how the line between the two concepts can be fuzzy, as exemplified in Baudrillard's concept of the ideological genesis of needs (1981, pp. 63–68). Were he alive today he might, as evidence of his stance, point to the claims made about the goodness of bottled water versus that drawn from the (properly) treated municipal supply. To drink water is to fulfill a need, but to drink commercially bottled water hyped as superior in flavor to tap water is to fulfill a want, a desire for water that tastes better than what flows from local taps.[6]

A discussion of leisure and consumption must be concerned primarily with wants rather than needs, unless of course, the needs are commercially transformed into wants. We need food, but it is leisure to meet this need at an upscale restaurant. Put otherwise some people *want* to meet this need in this way, though they do not have to do so simply to survive. Patronizing such a restaurant is a recognized leisure activity, albeit not one for all people. Nonetheless it stretches the imagination to regard the transformed need of drinking bottled water as a leisure activity. Wearing a designer coat could be viewed in the same light, as nonleisure, even while a much cheaper coat might keep its owner just as warm. Or the handy man or woman of the house wants to buy a recently advertised, good quality screwdriver with which to make obligatory repairs there, an activity not seen as leisure. In short, some needs are commercially transformed into wants that result in leisure, whereas other such transformations do not have this outcome.

So most of what concerns the field of leisure and consumption are directly created wants, desires that never were founded on one or more basic needs. A good deal of research, and no small amount of invective (e.g. Lefevbre, 1991; Barber, 2007), have focused on the tendency in the world of manufacture and sales to create these kinds of wants in possible buyers. At times this has been accomplished by sensitizing people through advertising and in-person sales pitches to wants they may not realize they have. There is no shortage of examples: listing the threats that call for a security service, the benefits to health of a certain herb, and the advantages of a particular savings account.

At other times, potential buyers know what they need, but commercial interests believe they should be pushed into buying a certain need-fulfilling good or service. We considered this process in Chapter 2 under the heading of technology and consumption. What is to prevent a seller from arousing in early adopters of, for example, a new kind of computer software their need for prestige, in this case that of possessing an unusual and conceivably valuable, useful good? The seller might also try to enhance sales by pointing out that early adopters may be sought out as sources of information about the product, adding to the prestige of ownership that of expert. In this scenario the early adopters know of their need for prestige, a sentiment the shrewd seller plays on.

The question all this raises for this chapter on leisure and consumption is just how forceful or demanding are these created needs? Do they amount to obligations that must be met? In line with our definition of leisure, free-time activities, to be regarded as such, may not be coerced. On the one hand, meeting some created needs might well be defined by some buyers as obligatory, among them, containing threats to security and gaining benefits to health. Given this outlook these people, were they to buy the service or product, would seem to be engaging in obligatory consumption. On the other hand, many a potential early adopter, though possibly enamored of the newfound identity of pioneering owner of a new technology and expert on its use and properties, might not, however, see this as obligation. Such an outlook would preserve the leisure basis for this sort of consumption, even though the purchase was motivated, in part, by created needs.

The prosumer

Alvin Toffler (1980) coined the neologism 'prosumer' to identify those consumers who help experts design certain products and services eventually to be created for people like them. The prosumer then consumes these products and services as part of the process of leisure customization (Godbey, 2004), which was described in Chapter 1. Toffler stressed the consultative side of 'prosumption' and customization by observing how would-be consumers were increasingly being asked what they wanted in a particular good or service. This proactive approach in the service and manufacturing sectors to meeting consumer wants, including needs transformed into wants, was seen as a new and effective avenue leading to growth in sales. Godbey's observations suggest, approximately 25 years later, that prosumers and their prosumption are alive and well, with both being especially important in the domain of leisure.

Toffler's prosumer extends the earlier ideas of Lefevbre, Baudrillard, and de Certeau in showing one concrete, common way by which the creative consumer can gain some control over his own consumptive destiny in the marketplace. These three thinkers stand out for the emphasis they place on social interaction between designers of various goods and services, on the one hand, and various sorts of buyers of those goods and services, on the other. If everything goes according to plan it is a happy arrangement for all concerned: consumers get what they believe are good products and designers as well as their manufacturers and service providers (strategies in de Certeau's theory) make a profit.

Laura Holson (2008) explains how this process works at LG Electronics, a company that makes, among many other products, mobile phones:

> LG electronics begins by asking focus groups to keep a journal, jotting down feelings about features they like most. Participants can call a toll-free number to share their emotions about the phone they are

testing. And sometimes they are asked to draw pictures that represent their mood when they hold the phone. 'Our job is to be behaviorists and psychologists,' said one LG vice president.

Holson writes that LG executives frequently attend home and design shows, hoping to spot relevant trends in popular culture that bear on their products. Today, in a highly competitive market, the mission of the company's marketing team is to identify the needs and wants of real people, some of whom are then recruited to LG's business plan as pro-sumers. In a different area of leisure, hobbyist quilters 'have been driving the current development of the quilting industry, and have used their checkbooks to nod approval to advances in quilting equipment, fabric development, shows/contests, quilt-related tours and cruises, extensive classes, and books and magazines that began in the mid-1970s' (Stalp, 2007, p. 6).

Shopping and community

What does all this mean for community? For most people much of the time, leisure-oriented shopping is casual and within their control. It is also more active and social than some other casual leisure activities. Thus here is where the communal side of shopping becomes most evi-dent, for it appears that people shopping in their leisure often do so in the company of others who enjoy the same activity.[7] Such shopping is likely to be with friends or relatives, if not both, but however, rarely with strangers.

But even here we encounter at least one major exception: *direct selling*. The Direct Selling Association (DSA) defines it as:

> the sale of a consumer product or service, person-to-person, away from a fixed retail location. These products and services are mar-keted to customers by independent salespeople. Depending on the company, the salespeople may be called distributors, representa-tives, consultants or various other titles. Products are sold primarily through in-home product demonstrations, parties and one-on-one selling.
>
> (www.dsa.org, retrieved 16 November 2008)

The DSA says that recent marketing innovations, including use of the Internet (e.g., 'DirectBuy') and mall kiosks, though currently less preva-lent than traditional direct selling approaches, are gradually becoming more common. Direct selling, this group maintains, is a good way to meet and socialize with people.

In a get-together that is almost exclusively female, a woman organizes in her own home a direct-sales party the centerpiece of which is a set of products she hopes to sell to invited guests and their interested friends and relatives. In the casual leisure atmosphere of such events, some food and drink is served as those present talk socially among themselves while examining and sometimes seeing demonstrations of commodities they may want to buy. The organizer receives a prearranged share of the profit from such sales, with the remainder going to the manufacturer of the products being sold. Although the DSA traces the history of direct selling to ancient times, it is possible that Tupperware (house wares) started the modern direct-sales trend. At least Zoe Brennan (2007) says that Tupperware parties were being organized even before 1951, the year the company began to rely exclusively on this way of selling. The Tupperware approach is now emulated by Shaklee (health care products), Pampered Chef (kitchen ware), Fifth Avenue Jewelry (jewelry), among several others. This was the sales model Merl Storr (2003) observed in her study of Ann Summers Parties (lingerie and sex toys) in Britain, which are clear occasions for direct-sales shopping as leisure.

Shopping, it is obvious, generates revenue for shopkeepers, some of which they may use as economic capital for advancing their businesses. But does shopping also generate social capital and thereby benefit the larger social community in some way? Social capital refers to connections among individuals, as manifested in social networks, trustworthiness, acts motivated by the norm of reciprocity, and the like. The concept is used by analogy to the concepts of human capital and economic capital (e.g., natural resources, financial resources) to emphasize that human groups of all kinds also benefit from and advance their interests according to the salutary interconnectivity of their members (Putnam, 2000).

Social capital is founded on the sustained interconnectivity of particular members of the community who until they met were strangers. Such a resource is built from links made among people who, before they joined a particular group or community project, were unknown to other participants in it. But shopping as leisure appears unlikely to generate social capital by bringing together members of the community who otherwise have no ties to each other, though it clearly fosters, in its own way, social relations with friends and relatives. So, shopping may be an occasion for mother and daughter to spend some time together, for two neighbors to go shopping for shrubbery with which to demarcate their mutual property line, or for a few staff employees to go to a specialty shop to buy small gifts for an office celebration.

Direct selling is a possible exception, in that direct-sales parties could conceivably nurture the development of local social capital because some of the guests are, at the outset, strangers to each other. Nevertheless trust and reciprocity can only be established over a period of time. Consequently these two conditions will fail to appear if new-found acquaintances at the direct-sales event cease interacting with one another once it is over. Social capital springs from, among other antecedents, relationships that have sufficient duration.

Citizen–consumer

The idea of the citizen as consumer, which has gained prominence in both the activist and the scholarly literatures on modern ethical consumption, links by way of the role of citizen the individual shopper with the larger society in which he lives. As Margaret Scammell notes, 'the act of consumption is becoming increasingly suffused with citizenship characteristics and considerations. . . . It is no longer possible to cut the deck neatly between citizenship and civic duty, on one side, and consumption and self interest, on the other' (Scammell, 2000, pp. 351–352). Evidence of the citizen–consumer is available in the socially conscious purchases in the marketplace made by, among others, environmentalists, anti-globalists, and members of the simple living movement. Their ideological positions and the purchases they motivate demonstrate a commitment to the local community and its larger society as well as the emergence of a new consumer subculture.

Nevertheless the idea is full of contradictions, which Josée Johnston has endeavored to clarify. She says that:

> while consumerism maximizes individual self-interest though commodity choice, the citizen–commons ideal prioritizes the collective good, which means that individual self-interest and pleasure can be trumped in the interest of improving sustainability or access to the commons. In short, citizenship struggles to reclaim the commons are collective, needs-oriented, and emphasize responsibility to ensure the survival and well being of others – human and non-human.
>
> (Johnston, 2008, pp. 243–244)

Johnston goes on to construct the consumer and the citizen as an interrelated pair of ideal-types, which may be used in comparative research (she conducted hers on the Whole Foods Market, an American corporation).The elements of these two types consist of the following:

> At the level of culture, an ideology of consumerism emphasizes the maximization of individual choice and variety, whereas citizenship encourages the bracketing of self-interest and the restriction of choice in the interest of collective solutions to achieve social justice and ecological integrity – in other words, to reclaim and preserve the commons. At the level of political-economy, consumerism links consumption to enhanced social status, as well as the maximization of one's own class status and well-being through consumption.... In the third realm of political ecology, consumerism supports the conservation of nature through consumption, buying more 'green' products in substitution for regular products (e.g., environmentally friendly disposable diapers), or buying products and services that construct a personal communion with nature (e.g., rainforest jungle tours or rainforest themed breakfast cereal). In contrast, a commons-focused ideology of citizenship advances disengagement with consumerism and reduced consumption, drawing from the political-ecological insight that current consumption levels of affluent populations are unsustainable and draw from the commons of peripheral regions.
>
> (Johnston, 2008, pp. 247–248)

Johnston's goal in conducting her ideal-type comparison is not to present consumer and citizen as two different empirical subjects existing as separate and isolated categories. Rather her aim is to fashion a heuristic tool capable of showing how consumerism and citizenship represent two contrasting explanatory frameworks with different normative ends. These ends, within the citizen–consumer hybrid, are irreconcilable, she holds.

Peter Taylor-Gooby (2006) is, in effect, working from Johnston's third realm of political ecology when he discusses the shift in Britain away from direct state management of the financial services market toward supporting citizen–consumers and their capacity to make educated choices in the marketplace. Taylor-Gooby quotes Christine Farnish, one-time FSA (Financial Services Agency, U.K.) Director of Consumer Relations, in her description of the role of that agency: 'to ensure that consumers are provided with the information they need, not only to understand what's going on, but also to help them make informed decisions about what they should do.' Regulation is no longer understood in Britain as a government-led process whereby government seeks to make the world safe for naïve and ill-informed buyers of goods and services. Instead consumers are now held to be knowledgeable and discriminating, able to take responsibility for their choices, though they may need first to be informed.

This redirection of responsibility to individuals puts human agency back in their hands. In addition this change squares well with the theoretic slant of this chapter, wherein shopping given its agency and limited choice, fits comfortably with the search for leisure, itself an underlying motive for such activity. We return to the citizen–consumer toward the end of Chapter 6.

Deviant consumption

Most of what we have considered to this point in this book has borne on conventional leisure and the diverse consumptive practices related to it. Still, the field of deviant leisure was introduced in Chapter 1 because some deviants, too, buy goods and services related to their immoral activities. This is an uncommon subject in both leisure studies and consumer studies, however, though that having been said, still one of considerable importance. Thus, to the extent the deviance is intolerable, it is of concern to law enforcement agencies. And since all deviance – tolerable and intolerable – is by definition in violation of certain moral norms, it is of greater or lesser concern for the conventional part of the community. Finally some deviance is big business, including excesses in gambling and the use of alcohol as well as the consumption of illicit drugs. All this justifies looking at some kinds of deviance as special varieties of shopping, where socially questionable goods and services are purchased.

It appears that most deviance and consumer-related activity is leisure rather than obligation. Paying for the services of a prostitute, purchasing bootleg whiskey, and buying a membership in a deviant group (e.g., white supremacists, Communist Party [where considered deviant], sadomasochist club, nudist resort) are done to facilitate free-time activity. By contrast, when an addict buys illicit drugs or an alcoholic buys legal (or illegal) liquor, such activity is coerced by compulsion, a state of mind classifiable in leisure studies as obligation. Most deviant shopping, however, does not appear to be of this forced nature.

Whether done for leisure or for obligation, some kinds of shopping defined by the community as deviant have a long history. Prostitution, drug and alcohol use, and deviant science (e.g., buying the services of a palm reader or crystal gazer, Stebbins, 1996c, p. 233) number among the most infamous examples. Julie Hardwick writes that such deviance was evident in 17th -century France.

> Yet, men who drank too much, who borrowed or lent money indiscriminately, or gambled more than they could afford discovered they had damaged their reputations as well as their credit. As a result, the costs of male sociability were a frequent subject of contention between spouses.
>
> (Hardwick, 2008, p. 461)

These breaches in public morality had to do, in one way or another, with consumption, albeit of the deviant variety.

Some consumers of deviant goods and services need to be knowledgeable about what they buy. True, for such hedonic, casual leisure activities as deviant sex, excessive playing of unskilled games of chance, and betting on illegal animal and bird fights, this knowledge is minimal, consisting primarily of where to find prostitutes, pornographic films, gambling establishments, illicit liquor, fighting animals, and the like. But the complexity of knowledge increases substantially in the deviant, serious leisure pursuits. Here, for instance, participants must know the many conventions that guide behavior at nudist parks (if nothing else, to help determine whether to buy a membership), the rules of playing face-to-face poker (skilled gambling), and the ways to cross-dress so as to pass as a member of the opposite sex and gain a sexual thrill while doing this.[8] The latter requires considerable knowledge of what to purchase in the way of clothing, make-up (for aspiring females), and accoutrements (e.g., purses, watches, necklaces) (Woodhouse, 1989). In the shady realm of tolerable deviance – poetically referred to by some as 'purple leisure' (e.g., Curtis, 1988) – men not women seem to be the most common shoppers, if for no other reason than the first are more likely to deviate.

Distance shopping is a thriving business in some areas of deviance. Online purchases of pornographic material and experiences constitute a lively example. Gambling online, when excessive, is now a well-established form of deviant leisure (see Moubarac, Gupta, and Martin, 2007, p. 522 for a short discussion of the problem online poker player). In what amounts to a sort of catalogue-based, distance shopping, some nudists and naturists peruse directories of resorts, beaches, clubs, cruises, events, and similar venues where nakedness is *de rigueur*. This may be either window-shopping, in effect, or initial exploration prior to making a purchase. The same may be said for the opportunities to engage in (buy services by way of) sex tourism, though in this instance, the opportunities promoted are found on the Internet (Kohm and Selwood, 1998).

The contexts of shopping for deviant purchases vary widely, from seedy poker backrooms, pornography shops, and points of sale for illicit booze to luxurious gambling casinos, posh nudist cruise ships, and high-class lounges where excessive alcohol consumption occasionally occurs. It appears that the higher the level of tolerance of the deviance in question the greater the likelihood it will be conducted in at least reasonably respectable circumstances. Notwithstanding the handful of up-market contexts most deviant commerce is carried out in quite ordinary, if not unsavory, surroundings. This condition, alone, sharply distinguishes conventional and deviant shopping.

As for creating wants the hedonic appeal of much deviant leisure indicates nothing much need be done to stir interest in searching for enjoyable goods and services. Moreover people and establishments hoping to attract patrons to deviant pursuits must, of necessity, keep a low profile, accomplished in part by only discretely calling attention to their existence. Instead it is generally left to the would-be buyer, fired by powerful desires, to find the means and places to meet them. As exceptions the world of gambling and, to a lesser extent, that of sex tourism (Oppermann, McKinley, and Chon, 1998) do publically promote their 'wares' and, in this manner, hope to stimulate consumer want to find leisure in them. The forces of initiative and human agency are highly evident here.

I know of no documented evidence of prosumption in the world of deviant activities, though large casinos and some nudist organizations might conceivably engage in it. And it is doubtful that citizen–consumer activity and civic responsibility exist anywhere in this distinctly self-centered area of life. It is, however, another story with direct selling. Acquisition of illicit liquor is typically accomplished on a direct-sales basis, as are most transactions between prostitutes and their johns. Informal gambling sessions, those arranged among friends or networks of friends and acquaintances and held away from commercial gambling establishments, may also be qualified as direct selling. In these gatherings, through wagers, participants pay money to play, to win, and to lose.

Deviant leisure in its consumptive phase is, it appears, rarely capable of generating any significant social capital. Thus, though strangers initially, men meet each other at an informal poker session held in a backroom of a bar or restaurant, where they pay (consume) as they play.

I argued earlier that direct selling is the only way that shopping sometimes generates social capital, an observation that gets further nuanced
when applied to the sphere of deviance. Yes, the men at a poker game
or an informal sport-betting session are directly meeting strangers and
exchanging money with them, possibly resulting in a few enduring
community linkages. But direct sales of illicit liquor and the services
of prostitutes are generally too fleeting and subterranean to amount
to create social capital of any kind. In brief, even direct selling of
deviant goods and services by no means always generates this community resource, which is on the whole, an uncommon consequence of all
shopping.

Conclusions

The goal of this chapter has been to identify and discuss several of
the key forces bearing on why people shop, what they buy, when
they do shop, and why they make their purchases. To this end, we
have looked at human agency and its constraints, historical trends,
motives underlying shopping (obligation, leisure), and knowledge of
goods and services. We also considered the way participation in the
marketplace is gendered, the effects of distance on consumption, and
the context of shops and shopping. Creating wants, prosumption, and
shopping to meet deviant interests were also covered. Moreover, since
shopping cannot be treated of in isolation from the wider community, we examined the extent to which it generates social capital and
the impact modern citizen–consumers can have on it through diverse
activist roles.

I make no pretense in this chapter of having covered all the key
forces. Many of the ones included in it have come up for scholarly
consideration, sometimes with notable ferment (e.g., creation of wants,
underlying motives), sometimes with minimal fanfare (e.g., deviant
consumption, human agency). How many other forces there are that
we have yet to discover remains to be seen. We will probably come
upon a number of them as research continues in this field, aided by
the fact that consumption studies is a decidedly interdisciplinary field.
This quality encourages exploration of its central focus from many
different angles and increases the chance of discovering new forces
that will help further explain Phase One of consumption. That is, the
rich array of concepts and theories found in each of the component
fields of this inter-discipline should be studied for their implications for
Phase One. Some of this has already occurred, as we have seen in, for

instance, gender and consumption, historical consumptive trends, and the influence of the citizen–consumer.

Following this exploratory procedure the next chapter will mine the field of leisure studies, a heretofore little-used angle from which to understand consumption, using its conceptual framework to illuminate Phase Two of the consumption process.

5
Phase Two: Consuming the Purchase

*One reasonably crisp way of distinguishing the two principal concepts of this book is to observe that the end of consumption is to **have** something, to possess it, whereas the end of leisure is to **do** something, to engage in an activity.* This proposition stands as one of the foundational observations of this book. Be that as it may we have already seen where consumption and leisure are so closely aligned as to make it impossible to distinguish the two, as seen in the examples of chasing down a rare coin and buying a fine old violin. The process of acquiring such items is seen by the collector and the musician as every bit a part of their serious leisure. Nevertheless such situations are exceptions to the proposition just presented. Accordingly, Chapter 4 was centered on the acquisition and possession of goods and services. The present chapter revolves around doing something with the goods and services the consumer now has.

But bear in mind that moving from Phase One to Phase Two is a fluid process, in that in making a purchase we always anticipate using it in some way. That is, the consumptive reality experienced by consumers would appear to be less cleanly demarcated according to the phases and types than our model may suggest. Consonant with this observation is the fact that both initiatory and facilitative consumption, which are discussed in this chapter, actually start in Phase One, while finding their most profound meaning in Phase Two.

Let us return to Belk's definition of consumption presented in Chapter 1. For him consumption 'consists of activities potentially leading to and actually following from the acquisition of a good or service by those engaging in such activities' (Belk, 2007, p. 737). In the present chapter we are concerned primarily with a main class of the activities actually following from the acquisition of a good or service. That is, interest here centers chiefly on leisure and leisure-like activities and less on those goods and services purchased to meet work and nonwork obligations. In obligatory activity people often buy goods and services intended to mitigate or facilitate meeting an obligation. For example, an assembly line worker buys an iPod to play music he hopes will lighten his hours on the job. An office employee purchases a fan to help cool her work station, exposed as it is to the hot summer sun.

This chapter examines, in much greater detail, two main consumptive processes that comprise Phase Two. They were briefly introduced in Chapter 2 as initiatory consumption and facilitative consumption. A section on nonconsumptive leisure follows. It shows, among other things, the extent to which leisure and consumption are separate worlds. Then we consider three issues that come onstage in Phase Two: leisure uncontrollability and selfishness; consumption, happiness, and positiveness; and authenticity and confidence.

Initiatory consumption

In initiatory consumption people quickly consume, in one way or another, what they purchase, what they have. In other words, they do what they intended with the purchased item either immediately or reasonably soon afterward. There are plenty of examples: the child buys a candy bar then eats it, a woman buys a theater ticket then watches a play, and a man buys a car then drives off in it. Or consider physicians, building contractors, and others, who must purchase expensive insurance to protect against suits arising from injury and alleged malpractice. The proximal effect of such a purchase is an instant sense of protection against these threats. In fact they may never, or only rarely, have to make a claim on their policy. As a final example, take the athlete who pays a physical therapist to work out a wrenched shoulder or rehabilitate a broken ankle. Here, too, the palliative effects of the therapy, as expected by both client and professional, are quickly evident. In the language just set out, all these buyers now have something of which they make more or less immediate use in some sort of subsequent activity. In other words, the physician and building contractor now work at ease, possible suits being covered by insurance, while the athlete now has a somewhat greater range of motion thanks to the ministrations of the therapist.

The main interest of this chapter lies in leisure-related initiatory consumption, those goods and services leading to participation in one or more free-time activities. Two of the examples presented in the previous paragraph lead to leisure – buying a candy bar and a theater ticket.

Buying a car is more difficult to classify. It depends on whether the transaction is defined by the buyer as meeting a disagreeable obligation (Jack's old car has become irreparable, forcing him to acquire another), a want to be satisfied (Ellen has long yearned for a BMW), or an agreeable obligation (Jane needs a car for work, but she adores driving the most recent models).

We looked at conspicuous consumption from this angle in Chapter 2. Veblen observed that 'conspicuous consumption of valuable goods is a means of reputability to the gentleman of leisure.' We saw that an expensive gift, an extravagant holiday somewhere, a sumptuous feast for friends and associates, must first be bought (conspicuously consumed) before they may be enjoyed (hedonically consumed). Aspects of primitive conspicuous consumption may be understood as initiatory consumption, notably the feasts and gifts. Obviously, however, the burning of possessions cannot be conceived of in this way. Indeed, in line with our definition, it fails to qualify as consumption of any kind. To the extent that reciprocity guides all this, obligation enters the exchange, raising the question of whether reciprocity in the *Potlatch* and the *Kula* is experienced as agreeable or disagreeable.

Leisure-related initiatory consumption may be motivated by a search for either casual leisure or its serious cousin (examples of both are presented later in this chapter). As for project-based leisure's relationship to this consumptive process, it appears to occur in one area only. This is the area of tourist projects, wherein the next step for the buyer of the air or boat ticket that launches the tour is to embark on it. Otherwise, from what we know at present about project-based leisure, it spawns only facilitative consumption.

The difference in initiatory consumption for casual leisure *vis-à-vis* its serious counterpart is evident in the following illustrations. Typically we purchase a candy bar for the immediate sensual pleasure its ingestion can provide. Theater tickets, however, may have a more complex meaning. If the buyer is seeking entertainment then, assuming the play is presented well, she will experience casual leisure. If she is a drama buff the initiatory consumptive act of purchasing a ticket enables her to spend some free time in her liberal arts hobby. In this scenario the distinction between having and doing is clear. Both theater goers *have* purchased a ticket, but they *do* different activities with it (experience entertainment vs. pursue a hobby).

In sum, even in initiatory consumption, the activities immediately following on buying or renting something may well be more profound than the comparatively superficial act of eating a candy bar. The literature reviewed in Chapter 3 suggests, largely simplistically, that it is otherwise, mostly making this point by denouncing the demoralizing consequences of overconsumption. In a way, this attack acknowledges the existence of our Phase Two, since something is said to happen after buying a good or service. But the complexities of initiatory consumption noted throughout this section have mostly gone unnoticed in this literature. Unnoticed as well is the enormous import for consumers and the society in which they live that springs from the vastly different roles consumption plays in casual and serious leisure. The serious leisure perspective set out in Chapter 1 helps explain this difference, while the next section describes it in consumptive terms.

Facilitative consumption

In facilitative consumption the acquired good or service only sets in motion a set of activities, which when completed, enabling the purchaser to use it in a more involved and enduring way than immediate consumption. Although much of facilitative consumption is related to serious leisure, it does occur in work and nonwork obligation as well. In general these purchases enable pursuit of a skilled, knowledge- and experience-based activity.

Turning first to work, some people, for instance, face the job-related requirement that they take a particular continuing education course designed to raise their occupational qualifications. To the extent that they oppose such mandated education, but nevertheless register for it, they engage in work-based facilitative consumption, accomplished by paying the course fee. Related costs of books, parking, transportation, and the like may also be seen as forced consumption. Now, it happens on occasion that employers pay for such personal betterment. In this case these continuing education students are not consumers, by our definition, even if some would still rather not take the course, which they see as a disagreeable obligation.

It is possible, too, that purchase for work of tools, manuals, equipment, and similar aids is understood this way, as necessary evils that, though they enhance performance there, still constitute expenses the worker would prefer to avoid. How many workers have been driven by the seemingly endless changes in computer software and hardware to upgrade both more frequently than they would like? Yet they regard these purchases as indispensable for staying in easy and efficient contact with associates in their field or their work organization or for effectively carrying out their occupational functions.

The foregoing has dwelt on work activities of which the workers are not much enamored, even if those activities require a significant level of skill, knowledge, and experience. But when conditions are right for the worker, such activities may become highly attractive. In effect they become a kind of serious leisure, albeit one pursued as a livelihood, as devotee work.

Occupational devotion

Such work has been conceptualized as 'devotee work,' involvement in which is fired by 'occupational devotion' (Stebbins, 2004a). There is in the West a small proportion of the working population who find it difficult to separate their work and leisure. These workers, for whom the line between the two domains is blurred, do rely on their work as a livelihood, but nevertheless are also 'occupational devotees.' That is, they feel a powerful occupational devotion, or strong, positive attachment to a form of self-enhancing work, where the sense of achievement is high and the core activity is endowed with such intense appeal that the line between this work and leisure is virtually erased. Further, it is by way of the core activity of their work that devotees realize a unique combination of, what are for them, strongly seated cultural values (Williams, 2000, p. 146): success, achievement, freedom of action, individual personality, and activity (being involved in something). Other categories of workers may also be animated by some, even all, of these values, but fail for various reasons to realize them in gainful employment.

Occupational devotees turn up chiefly, though not exclusively, in four areas of the economy, providing their work here is, at most, only lightly bureaucratized: certain small businesses, the skilled trades, the consulting and counseling occupations, and the public- and client-centered professions. Public-centered professions are found in the arts, sports, scientific, and entertainment fields, while those that are client-centered abound in such fields as law, teaching, accounting, and medicine (Stebbins, 1992, p. 22). It is assumed in all this that the work and its core activity to which people become devoted carries with it a respectable personal and social identity within their reference groups, since it would be difficult, if not impossible, to be devoted to work that those groups regarded with scorn. Still, positive identification with the job is not a defining condition of occupational devotion, since such identification can develop for other reasons, including high salary, prestigious employer, and advanced educational qualifications.

The fact of devotee work for some people and its possibility for others signals that work, as one of life's domains, may be positive. Granted, most workers are not fortunate enough to find such work. For those who do find it, the work meets six criteria (Stebbins, 2004a, p. 9). To generate occupational devotion:

(1) The valued core activity must be profound; to perform its acceptability requires substantial skill, knowledge, or experience or a combination of two or three of these.

(2) The core must offer significant variety.

(3) The core must also offer significant opportunity for creative or innovative work, as a valued expression of individual personality. The adjectives 'creative' and 'innovative' stress that the undertaking results in something new or different, showing imagination and application of routine skill or knowledge. That is, boredom is likely to develop only after the onset of fatigue experienced from long hours on the job, a point at which significant creativity and innovation are no longer possible.

(4) The would-be devotee must have reasonable control over the amount and disposition of time put into the occupation (the value of freedom of action), such that he can prevent it from becoming a burden. Medium and large bureaucracies have tended to subvert this criterion. For, in interest of the survival and development of their organization, managers have felt they must deny their nonunionized employees this freedom, and force them to accept stiff deadlines and heavy workloads. But no activity, be it leisure or work, is so appealing that it invites unlimited participation during all waking hours.

(5) The would-be devotee must have both an aptitude and a taste for the work in question. This is, in part, a case of one man's meat being another man's poison. John finds great fulfillment in being a physician, an occupation that holds little appeal for Jane who, instead, adores being a lawyer (work John finds unappealing).

(6) The devotees must work in a physical and social milieu that encourages them to pursue often and without significant constraint the core activity. This includes avoidance of excessive paperwork, caseloads, class sizes, market demands, and the like.

Sounds ideal, if not idealistic, but in fact occupations and work roles exist that meet these criteria. In today's climate of occupational deskilling, over-bureaucratization, and similar impediments to fulfilling core activity at work, many people find it difficult to locate or arrange devotee employment. The six criteria just listed also characterize serious leisure (see Stebbins, 2004a, Chapter 4), which gives further substance to the claim being put forward here that such leisure and devotee work occupy a great deal of common ground.

Facilitative consumption in devotee work sometimes becomes positively associated with this leisure-like activity as a vehicle for enhancing the experience. Consider the occupational devotee required by the boss to take a specialized course through continuing education that the former is eager to enroll in, precisely because the knowledge gained there will improve performance in a highly attractive job. And this even if the devotee has to pay all expenses. As observed above some (nondevotee) workers detest having to buy computer hardware and software. Yet, writers, for example, enamored of their occupation might well see a recently purchased, updated computer and accompanying software as a major improvement to getting their ideas 'on paper' more efficiently and effectively compared with what they were using. Could not acquisition of a new, substantially improved tool elevate significantly the attractiveness of a craftsman's work, whose core activities he or she is already devoted to?

Additionally purchased goods that enable buyers to perform better in their devotee work role may acquire special positive meaning. Being associated with the elevated performance, perhaps even seen as an indispensable part of it, brings the good in question close to the heart of the worker. Kevin Jonathan Lee, an American luthier, discusses the ins and outs of buying a violin, ranging from the cheapest to the most expensive, and the emotional boost that can come from playing on a fine instrument:

> Special note: If you have the means, buy the finest violin you can, regardless of your current performing level. It's so much easier playing on a fine violin, and you'll enjoy the experience that much better right from the beginning.... As the price increases, the names (and usually the quality) improves until true master instruments are reached (though by today's standards, the ones in this price range are not up to the challenge of solo performances in large concert halls with large symphonies. Yet, as always, there are exceptions and many professionals have found their 'companions for life' within this price range).
>
> (http://www.kevinleeluthier.com, 'What to look for (and look out for) when purchasing a violin,' retrieved 21 December 2008)

Of course these observations hold not only for professional musicians but also for their serious amateur colleagues.

Serious leisure

This form of leisure was described in Chapter 1. There it was defined as the systematic pursuit of an amateur, hobbyist, or volunteer activity sufficiently substantial, interesting, and fulfilling for the participant to find a (leisure) career there acquiring and expressing a combination of its special skills, knowledge, and experience. What remains to be done in the present section is to show how Phase Two of consumption, when done in the name of serious leisure, has its special properties.

Let us start with an example that has been following us through this book: amateur violinists. If they are to play at all, they must first rent or purchase a violin – an act of acquisition, of having.[1] Yet their most profound leisure experience is competently and artistically playing music and regular practice to accomplish this, all of which costs nothing, though, obviously, it is certainly facilitated by playing on the acquired instrument, the higher its quality the better (a consumer product). Moreover, for amateurs, this profound leisure experience may be further facilitated by paying for music lessons and buying public transit tickets to travel to their teacher's studio.

Many serious leisure pursuits require one or more prerequisite purchases, but here, as in devotee work, participants accent the highly appealing core activities of their leisure. The purchased goods that enable their buyers to perform better in this form of leisure may also acquire special positive meaning. Being associated with the elevated performance, perhaps even seen as an indispensable part of it, brings the good in question close to the heart of the worker. How much more positively exciting, or meaningful, is the bus ride to a music lesson for the committed student violinist than one of equivalent distance to, say, work for the nondevotee employee. The first ride is filled with anticipation of learning new technique, perfecting that already learned, and hearing the teacher play with enviable expertise and emotion. Or consider the excitement of Agnes de Mille, a renowned American ballerina, during her days as a student and amateur:

> I woke blissfully on Thursday morning. 'Today I have a lesson with Miss Fredover.' Friday entailed disappointment since it introduced a whole week before the next private lesson. Three times during the first winter, Miss Fredova said 'very good' and I recorded the event duly in my diary. On those nights I drove home with a singing heart

and stood in the bedroom in the dark gripping the edge of my desk in excitement, so in love with dancing, so in love with her. 'Oh God,' I prayed, 'let me be like her. Let me be a fine dancer.'

(de Mille, 1952, p. 56)

As for the second ride it probably evokes such expectations as boredom, physical fatigue, and fractious relations with the boss or a coworker. Soon this ride becomes colored with negative meaning.

Let us return to the liberal arts hobbies of reading, many of which rely heavily on buying or borrowing books. Richard de Bury speaks lovingly of books, in a manner that demonstrates how consumptive serious leisure is facilitated in this type of activity and how the chief facilitator – books – becomes closely and emotionally associated with the development of knowledge, pursuit of truth, and experience of personal happiness.

> Moreover, since books are the aptest teachers, as the previous chapter assumes, it is fitting to bestow on them the honour and the affection that we owe to our teachers. In fine, since all men naturally desire to know, and since by means of books we can attain the knowledge of the ancients, which is to be desired beyond all riches, what man living according to nature would not feel the desire of books? And although we know that swine trample pearls under foot, the wise man will not therefore be deterred from gathering the pearls that lie before him. A library of wisdom, then, is more precious than all wealth, and all things that are desirable cannot be compared to it. Whoever therefore claims to be zealous of truth, of happiness, of wisdom or knowledge, aye, even of the faith, must needs become a lover of books.
>
> (de Bury, 1909, pp. 17–18)

At the time the present book is being written – during one of the world's worst economic crises – uses of serious leisure seldom discussed heretofore have come to light. Claire Miller (2008) describes how some people in the United States are coping with today's economic downturn, in part, by turning to craftwork. Stores that sell supplies for handmade goods are experiencing a boom of sorts, as an unusual number of customers have decided it is cheaper to make their Christmas gifts than buy the commercial equivalent. Some retailers believe that this trend is broader than the holiday season and that it includes discovery of a new way of relieving stress, no small part of which is triggered by the dreadful economy.

Of theoretic interest is whether these 'new' crafts enthusiasts are hobbyists who, for reasons such as those listed above, have stepped up their rate of participation. Or are they neophytes just starting a new crafting pastime (in this instance one might wonder about the quality of the gift)? A third possible explanation is whether the item created is intended as an expression of a hobby, as in either of the first two alternatives, or as a one-off project. These questions beg research.

Project-based leisure

Much of project-based leisure appears to be consumer based. Thus most, perhaps all, one-off projects require preliminary purchases, though not, however, of the momentous variety seen in the earlier illustration involving the expensive violin. The same may be said for the liberal arts projects, with the possible exception of constructing a genealogy. Although computer programs may be bought for this purpose, some people prepare their genealogies with little cost, as by writing and telephoning relatives and writing up their results by hand (Lambert, 1996). Finally activity participation seems to invariably involve purchase of equipment and travel services. Indeed getting to some of these activities, itself often a major expense, may be quite involved and most unpleasant (e.g., a long air trip to the base of Mt Everest). But this is still not a core activity of the sort described above in acquiring certain collectibles.

By contrast one-off volunteering projects, with one possible exception, can be qualified as nonconsumptive leisure. That is, unless we count as acquisitions the costs of transportation, clothing, and food borne by the volunteer while engaging in the altruistic activity and for which the festival, museum, or sporting organization offers no reimbursement. As an example consider the lot of those disaster volunteers who have to spend a great deal of their own money on meals, lodging, and transportation to get to the scene of a hurricane or oil spill and then to stay around to help ease the damage.

Much of the time occasional projects seem to require that participants buy something to bring them to fruition. Holding a surprise birthday party is bound to generate costs, as does decorating the house and garden for the Christmas season. Here, too, exceptions exist, among them, the research that some foresightful people do to buy a new car, which they carry out on the Internet and at certain dealers (this example assumes they are not coerced to make this purchase). Another exception is evident in those who envisage a major holiday in a far-away destination. Some of them plan for it by consulting the Internet, relevant print media, and possibly, knowledgeable friends and relatives. But here they are engaging in facilitative consumption that is simultaneously a core activity. For, to plan a tour is to imagine how it should unfold and to try to ensure that it will unfold as planned.

Project-based leisure seems to show better than the other two forms of leisure how leisure consumption actually bridges the two Phases. True this form has been covered in this section on facilitative consumption, and the examples just presented generally fit here. But the project of doing research to buy a car is most accurately classified as initiatory consumption, even while rather little money is spent doing it. The role of consumer knowledge, discussed in Phase One, is also evident in some leisure projects. In fact, gaining such knowledge can become a project in itself, as an aid to shopping, to buying certain goods or services.

Nonconsumptive leisure

To round out further the conceptualization of leisure and consumption, to provide further background, we must also look at nonconsumptive leisure. Such pursuits abound largely, though not exclusively, in casual and serious leisure. In nonconsumptive leisure, activities cost nothing, or at most, the costs are negligible. In consumption that is negligible only relatively small amounts of money are spent on the leisure activities in question.

I argued in the Chapter 1 following Roberts (1999, p. 179) and Kiewa (2003, p. 80), that in no way can all of leisure can be equated with consumption, even mass consumption. The vast area of nonconsumptive leisure gives substance to their claim, while showing in detail what is meant by the subtitle of this book. The latter depicts leisure and consumption as occupying common ground, while nonetheless being separate worlds.

Casual leisure

All eight types of casual leisure contain nonconsumptive activities. In play dabbling often occurs free of charge on borrowed equipment, be it a piano, tennis racket, or telescope. To buy such things merely to play around with – to use them as toys – would be unthinkable for most people. Daydreaming is mental dabbling, and it costs nothing. Turning to relaxation, it is certainly possible to spend substantial sums to lounge beside the pool at an opulent resort, drink fine wine and watch passers-by from a sidewalk café on Boulevard St Germain in Paris, or luxuriate with a massage at an upscale spa. But far more accessible for most people is relaxing without cost by taking an afternoon nap in the easy chair at home, casually strolling through a local park, or listening to favorite tunes while sunning on a community beach. Some entertainment is costly, whereas other forms are available at negligible expense. Watching television usually falls beyond this second group, however, since buying a set and paying for cable service make this practice rather dear. Yet, observing a busking musician, listening to the radio, and watching an air show may be accomplished with little or no money.

Sensory stimulation may be virtually, if not literally, free. This includes watching a vivid sunset or a flowing brook, having sex (not with a prostitute), listening to birds sing, and watching children play or the family dog chase a ball. Sociable conversation, unless inspired by expensive coffee or liquor, costs nothing or next to it. One celebrated genre – gossip – can in certain circles fill several hours a week. And much of casual volunteering costs nothing apart, possibly, from outlays for transportation and clothing. The examples of pleasurable aerobic activity presented in Chapter 1 may all be pursued without expense. Still this kind of leisure can be expensive, as for instance, when Wii Fit, PlayStation 2, and other video games are played while working out an exercise bike, treadmill, or similar device, all being equipment that must be bought directly or borrowed. The second is usually only possible after paying a fee or buying a membership.

Lisa Dunning, a California-licensed marriage and family therapist specializing in parent/child relational issues, describes some of the rewards to be gained from the casual leisure activity of watching children at play:

> You can learn a lot by watching children playing. There is no greater sight than to see children pretending to be superheroes or digging for treasure in a sandbox. To hear the laughter and see a child's excitement can bring a smile to any adult's face. An important part of a child's development is through play. A child's problem solving skills are developed through play. Children are able to work on relationships, build self-esteem and decrease their anger. In the course of a child's life adults usually teach children, but on the playground children can teach adults how to have fun.
>
> If you are having difficulty in your life and are in need of a break just take the time out of your busy unrelenting day to watch children

play. If you feel comfortable doing so, join in the fun. The stresses and strains of your daily life will just slip away.

(www. ambafrance-do.org/stress-management, retrieved 26 December 2008)

There is here both sensory stimulation (e.g., joy, talk, and movement of children) and relaxation – casual leisure without charge.

Serious leisure

Numerous activities in this form also qualify as being of negligible, if not nonexistent, cost, including much, if not all, of those classified as volunteering. Here, to engage in the core activities, the enthusiast need not acquire something expensive, whether by buying or renting it. The same is true for a variety of hobbies, among them, the liberal arts reading hobbies (e.g., exploring a genre of history or science), some collecting hobbies (e.g., leaves, seashells, insects), and some outdoor sports and activities (e.g., walking in nature, swimming in a lake, playing soccer or touch football).

Look more closely at volunteering. To participate in this form of leisure, many volunteers must travel somewhere to fill their role (e.g., at a hospital, playing field, community park, school function), and this might cost them money for petrol or transit fares. A volunteer might have to buy special clothing for the activity, as in warm apparel for waiting outside in winter to catch the bus or comfortable shoes for long periods of standing as a regular usher in a community concert hall. Perhaps there is need to purchase a manual or take a course. Money is paid out in these examples, but only infrequently or in small amounts. Most people would not find these Phase Two expenses prohibitive, such that the expenses might discourage them from participating in the activity in question. Generally speaking volunteering is low cost, unless the leisure sought in this vein is that of, say, volunteer tourism (Wearing, 2001) or volunteer missionary work. Disaster volunteering and volunteer search and rescue may also be costly, mainly for the reasons set out above where we discussed the first.

Next consider the collecting hobbies. Obviously there are hobbies of this nature where great sums of money are expended in acquiring collectibles, among them, collecting coins, paintings, old cars, musical instruments, and antique furniture. Besides buying the collectibles themselves, there may well be substantial costs associated with finding them and, for old cars, with transporting them to the collector's residence. Other items cost less than these, but are nonetheless far from free. Collectors of plates, dolls, posters, figurines, and the like may still pay significant sums for them, though often these collectors seem to shop for their collectibles while on a holiday rather than on a special trip expressly to buy them. Still plate and pottery collectors, when collecting far from home, may face some jaw-dropping shipping costs.

It is the collector of natural objects who is most likely to find nonconsumptive leisure. Of course this person must usually consume some petrol to get to the forest to collect leaves or fossils, the mountains to collect rocks or minerals, or the shore to gather shells or starfish. Furthermore some of these enthusiasts may conclude that going to these places is difficult enough to justify owning a four-wheel drive vehicle, which for them, removes this hobby from the category of nonconsumptive leisure. And some rock hounds may want a microscope to examine more closely what they have found. Collectors of natural objects of any kind may also need to buy special clothing to participate in their activities.

Consider the hobby of collecting sea shells, as presented by www.seashells.com (retrieved 26 December 2008):

> Walking along the beach picking up seashells and sea life has been enjoyed by millions of people throughout the world. There is nothing more heart warming then watching a young child run along the beach for hours on end excitedly pointing out all of the fabulous creatures and sea shells that can be found along the seashore, and gathering up seashells to listen into. With so many miles of shorelines throughout the United States and the world, all covered with nature's treasures, it would be difficult not to find this hobby enjoyable. Remember that our coastlines are an important part of the environment, so please try to leave an area as pristine as it was when you arrived, if not more so. Please do not litter and if you see someone else's trash gather and dispose of it properly. If you flip over a rock to look for specimens make sure to replace it when you are done looking. All of these tips will help to keep our shorelines a viable habitat for all.

This site goes on to describe how to identify and classify the shells that may be found on ocean shores, using the following headings: molluscs, sand dollars, star fish, sea urchins, sponges, crabs, sea fans, and 'other oddities' such as egg casings, sea horses, trigger fish, and mermaid purses. This sort of nonconsumptive leisure grows marginally more expensive with acquisition of one of the many guidebooks for identifying shells and possibly a camera (it may be disposable) for recording the collecting experience and the collectibles found.

Nonetheless, a case may be made for qualifying this kind of serious leisure as, in general, essentially nonconsumptive. Essentially nonconsumptive so long as it remains as collecting and does not evolve into, for instance, a making and tinkering hobby (e.g., polishing rocks) or amateur mineralogy (e.g., conducting descriptive analyses on the collected rocks according to established scientific theory and taxonomy). These latter-day types of serious leisure require specialized equipment. Therefore they lead the participant into the marketplace and the vast sphere of consumptive leisure.

What about the liberal arts reading hobbies? At its cheapest this hobby may be pursued by reading books, magazines, and similar sources at one or more libraries or by borrowing such material from friends, relatives, and others, if not both. No purchases in this approach, again with the possible exception of, in the main, the cost of transportation to reach the libraries or the other sources. Nevertheless the cost of liberal arts reading goes up to the extent that the hobbyist buys the reading material, or gathers the desired information from other outlets for which there is a charge. These outlets include lectures, museums, television documentaries, adult education courses, and the Internet, all of which may require the individual to pay a certain amount. Costs in the liberal arts hobbies begin to move beyond being negligible when these hobbyists must pay what they regard as a burdensome amount for any or all of a television set, Internet service, adult education, or books and magazines (say, because the library is inaccessible or inadequate).

Next, look at the 'nature challenge' sports and other hobbies pursued in one or more of six natural settings composed essentially of: (1) air, (2) water, (3) land, (4) animals (including birds and fish), (5) plants, and (6) ice or snow (sometimes both) (Stebbins, 2009c).[2] First, note however, that indoor versions of these, which occur in artificial settings, to the extent they exist, cost something – commonly an entry fee or a membership. This may also be true of those available out of doors, as found, for instance, in public or private parks and in a variety of specialized recreational facilities and services (e.g., alpine skiing, white-water rafting, rental canoeing, professionally guided back packing or mountain climbing, guided horseback riding). Some of these can be enormously expensive, as most every alpine skier would attest.

Yet, there are many activities that can be undertaken in nature settings that cost nothing other than, possibly, transportation and appropriate clothing. Some of these were listed early in this section. Thus, walking in nature includes hiking, cave exploration, and snowshoeing. Bird watching need not be overexpensive, while mushroom gathering may be done at low cost (i.e., that of a pail). Swimming in a lake or the sea can often be undertaken from a public or other noncommercial shore line without charge and, for some, allows the opportunity to snorkel, the device itself being cheaply available.

As with some of the aforementioned activities, the ones in this paragraph are subject to sometimes dramatic escalations in cost. Bird watching expenses mount quickly with the purchase of a sophisticated pair of binoculars, a set of bird books, and travel to different parts of the world to view unfamiliar species. Hiking, seemingly an invariably low-cost activity, is subject to heavier expenses incurred in buying boots suitable for mountain scrambling (hiking without technical equipment to mountain summits, ridges, and passes), clothing for a wide variety of weather conditions, and nowadays, maps and a global positioning instrument for effective passage in areas where trails are poorly marked or simply nonexistent. Most modern snowshoeing is performed on lightweight aluminum or neoprene frames, modeled in part, on the traditional wood and leather artifacts invented by the North American Indians. The traditional snowshoe is still commercially, albeit cheaply, available (negligible cost) compared with its modern counterpart. Either way, buying a pair of snowshoes is a long-term purchase, suggesting that this hobby is essentially nonconsumptive leisure.

Turning next to the various hobbyist sports and games pursued in artificial settings, we find many whose costs are negligible, among them darts, horseshoes, shuffleboard, croquet, and ping-pong. A set of well-balanced darts would constitute an initial and, at most, moderate expense that would not have to be repeated, however, for many years. The greatest cost of the others in this list, if any, is probably membership in a club or other organization that offers a horseshoe pit, shuffleboard court, or ping-pong table, where players may meet and compete with each other.

Games are, but with few exceptions, cheap hobbyist, nonconsumptive leisure. It costs a pittance to buy the board games of Monopoly, Parcheesi, backgammon, and checkers. Most chess sets are reasonably priced, though some of the ornate or luxurious models (e.g., with silver or ivory pieces) may cost a fortune. Card games, especially those based on the standard 52-card deck, are very low in price. The same holds for dominoes, Scrabble (though it helps to buy a Scrabble dictionary), charades, Pictionary, and the role-playing games (e.g., Dungeons and Dragons, Empire of the Petal Thorne). Where card games exit the sphere of nonconsumptive leisure is when they become vehicles for routine medium- and high-stakes gambling. Charades and the role-playing games are, in fact, literally free of cost, that is unless we take into account the expenses of transportation and those of refreshments before, during, and after a session of gaming.

Finally in the hobbyist field several making and tinkering activities may be pursued at negligible expense. For example, scrapbooking, origami, and many of the interlacing, interlocking, and knot-making activities may be done for the cost of the materials, generally miniscule. Paper, yarn, rope, and the like are not commonly dear. The same may be said for collage, *découpage*, and a great range of decorating pursuits. The latter include adorning small objects like toothpicks, thimbles, washers, and drinking glasses, to name a few. Creating mobiles and stencils also falls into this class of hobby. Understandably decorating with flowers, etching glass, and burning wood are typically more expensive pastimes, requiring equipment or resources whose costs accumulate. This observation also holds for hobbyist cooking, baking, brewing, candy-making, and wine-making.

Turning to amateurism there are here, as well, pursuits of negligible cost. Vocal music and stage drama are examples.[3] In the main the only art that is low cost is that of drawing and sketching (in ink, pencil, charcoal). All types of amateur writing are of negligible expense, once the writer acquires an acceptable computer, which for fiction and poetry would not normally even need an Internet hookup. Sport is generally costly. Even such activities as handball, volleyball, swimming, rowing, and water polo though cheap to participate in commonly require an entry fee for or a membership in a club or recreational facility. Amateur science lies almost uniformly beyond the realm of low cost, primarily because equipment is more or less expensive. History, when pursued as a library and documentary science, is the main exception.

Project-based leisure

Some projects are one-off or occasional undertakings that become hobbies if they come to be pursued regularly over many years. Yet, many people stop short of this progression. Thus, an inexpensive interlacing, interlocking, or knot-making project might be tried once and perhaps repeated occasionally over time, but not often enough to be considered serious leisure. Some family history projects can be of the same nature, providing expenses in archival work and contacting living relatives are minimized. Additionally some entertainment theater, if costs remain low, can be qualified as nonconsumptive leisure. Examples include preparing and presenting a skit, public talk, or puppet show.

Consider the leisure experienced in producing a detailed single-family history and the nature of the monetary costs commonly incurred.

In the narrower sense of the term, a family history is a biography of a single family over several generations, based on a tested genealogy and fleshed out with the fuller story of the family's place in society, the dramas of its achievements or failures and its acquisition or loss of wealth and rank.

Such a study mainly draws on oral history for the recent period and archival records for the period beyond living memory. Where an individual's own story is unknown, much can be inferred from other literature. For example, a single soldier's experiences can be inferred from the history of his military unit, or a migrant's journey can be described from the shipboard diary of a fellow traveler.

Family history can either be in the form of a printed document, electronic document or sound or video recording that preserves this history for future generations. The readers will expect it to describe where the family originated from, name the members of the family and state who they married.

Family Histories are often created as a memorial for the deceased and are written to be passed down to future generations. Some records that are used to create family histories are:

- Apprenticeship records.
- Baptism or christening records.
- Birth certificates.
- Cemetery records and tombstones.
- Census records.
- Coroner's reports.
- Death records.
- Diaries, personal letters, family Bibles, scrapbooks and ephemera.
- Directories – trade directories, street directories, telephone directories.
- Earlier family histories.
- Marriage records.
- Military records.
- Newspapers – both news items and advertisements.

- Property records and contemporary maps.
- Public records – Social Security records (in the U.S.), Poor Law records (in the U.K.), registers of electors.
- Tax records.
- Wills and probate records.

Today many people are using these old records to recover their family history. But most of these records include only technical details of a person's life, such as their birth date, whom they married, the jobs they did, and so forth, but they contain very little about the person themselves such as their likes, dislikes, hobbies, hopes and dreams.
(Wikipedia 'Family History,' retrieved 26 December 2008)

The main costs here are those of transportation to sources, photocopies of records, and use of postal and telephone services to contact relatives. If all this is done locally or regionally, the principle of negligible cost may very well be honored. But in visits to distant places expressly to create a family history (e.g., air ticket, car rental, room and board), expenses can quickly mount, becoming significantly costly facilitators for a hobby no longer classifiable as nonconsumptive leisure.

The occasional projects are highly diverse, and their full extent anywhere in the world is not yet known. Celebratory projects such as special birthdays, national holidays, and religious events, tend to require significant outlay. Hence they fail to qualify as nonconsumptive leisure. Indeed it may turn out, upon closer examination of this area of free time, that the occasional projects are always, or nearly always, substantially consumptive.

This section has been intentionally long and detailed. Necessarily so, for it constitutes, as it were, the evidence for our proposition that consumption and leisure occupy, in significant measure, separate worlds, that leisure and consumption are not an identity. This is an academic claim, however, and from a practical standpoint possibly one of scant interest. Yet, what is of practical interest is the fact that considerable leisure exists in wide variety, which people may pursue with little or no money but in which they may still find great fulfillment. The foregoing lists of nonconsumptive leisure – and they are by no means exhaustive – though perhaps of marginal interest in a scholarly book on leisure and consumption, are vital to fields bent on finding acceptable and absorbing leisure for youth (as an alternative to deviance), low-income groups, and other people who have some money though little of it remains after meeting basic needs.

Leisure uncontrollability and selfishness

From what was said in Chapter 1 about costs and rewards, it is evident why the desire to participate in the core amateur, hobbyist, or volunteer activity can become for some participants some of the time more or less *uncontrollable*. More precisely it energizes them to engage in the activity beyond the time or the money (if not both) available for it. As a professional violinist once counseled his daughter, 'Rachel, never marry an amateur violinist! He will want to play quartets all night' (from Bowen, 1935, p. 93). There seems to be an almost universal desire to upgrade: to own a better set of golf clubs, buy a more powerful telescope, take more dance lessons perhaps from a renowned (and consequently more expensive) professional, and so forth. And, yes, play more and more string quartets (doing so by purchasing more and more music). The same can be said for the hobbyist and volunteer pursuits.

Chances are therefore good that some serious leisure enthusiasts will be eager to spend more time at and money on the core activity than is likely to be countenanced by certain significant others who also makes demands on that time and money. The latter may conclude sooner or later that the enthusiast is more enamored of the core leisure activity than of, say, the partner or spouse.[4] Charges of selfishness may then not be far off.

Selfishness is the act of a self-seeker judged as selfish by the victim of that act (Stebbins, 1981). When we define an act as selfish, we make an imputation. This imputation is most commonly hurled at perceived self-seekers by their victims, where the self-seekers are felt to demonstrate a concern for their own welfare or advantage at the expense of or in disregard for those victims. The central thread running through the fabric of selfishness is exploitative unfairness – a kind of personal favoritism infecting the everyday affairs of many people in modern society. In comparing the three forms, it is evident that serious leisure is nearly always the most complicated and enduring of them and, for this reason, often takes up much more of a participant's time and money (Stebbins, 1995). Consequently it is the most likely to generate charges of selfishness.

I found in my research on serious leisure that attractive activity and selfishness are natural partners (Stebbins, 2001a, Chapter 4). Whereas some casual leisure and even project-based leisure can also be uncontrollable, the marginality hypothesis (see next paragraph) implies that such a proclivity is generally significantly stronger among serious leisure participants. Selfishness is an ethical question seldom raised in leisure studies.

Finally, I have argued over the years that amateurs, and sometimes even the activities they pursue, are marginal in society, for amateurs are neither dabblers (casual leisure) nor professionals (see also Stebbins, 1979). Moreover, studies of hobbyists and career volunteers show that they and some of their activities are just as marginal and for many of the same reasons (Stebbins, 1996b, 1998). At least two properties of serious leisure give substance to this hypothesis. One, although seemingly illogical according to common sense, is that serious leisure is characterized empirically by an important degree of positive commitment to a pursuit (Stebbins, 1992, pp. 51–52). This commitment is measured, among other ways, by the sizeable investments of time, energy, and sometimes money in the leisure made by its core devotees *vis-à-vis* its more ordinary participants (see Chapter 1 of this book). Two, serious leisure is pursued with noticeable intentness, with such passion that Erving Goffman (1963, pp. 144–145) once branded amateurs and hobbyists as the 'quietly disaffiliated.' People with such orientations toward their leisure are marginal compared with those who go in for the ever-popular activities characteristic of so much of casual leisure.

Would it be more valid to describe serious leisure devotees as fanatics rather than as uncontrollable? Fanatical is how they are sometimes described in the literature on event studies (for a review of this consumer field see Mackellar, 2006). Here fanatics are said to approach their leisure with excess and extreme passion, an idea that is, however, also loosely used in common parlance. In this sphere usage of the term suffers from a widespread lack of objectivity as well as a cultural bias that portrays fanatics as deranged. For this reason it has faced an uphill fight to gain acceptance as a scientific concept. Both fanatics and uncontrollable serious leisure participants tend to buy more goods and services related to their core activities than they can justify to people outside their hobby or amateur pursuit. This is why they need to be included in this book. Be that as it may, the term 'uncontrollable participant' seems preferable, if for no other reason than that it lacks the commonsensical baggage just attributed to 'fanatic.' Moreover the latter, although now also used figuratively, continues in common sense to refer as well to its original meaning of religious frenzy arising from being possessed by a god or a demon.

Consumption, happiness, and positiveness

As Samuel Butler observed over a century ago: 'It has been said that the love of money is the root of all evil. The want of money is so quite as truly' (Butler, 1917). If both the presence of money and its lack lead to evil, as this quotation attests, might it be that the stuff really does buy happiness? Or in the preferred language of this volume, might it buy goods and services that would increase the happiness of their purchaser? Many a popular book and magazine article trumpets an affirmative answer to this question, and then serves up advice on how this may be accomplished. But these are, in effect, only manuals; they are not studies backed by data showing that, whatever the means, money sometimes, or always, buys happiness. On the one hand, then, the manuals fail to offer convincing evidence that the formulae listed within lead to the promised land of wealth-based joy.

On the other hand, there *is* evidence demonstrating that certain non-material factors bear on the human happiness quotient, and that money may, under some conditions, actually hinder our pursuit of this emotion. For example, Andrew Oswald and Nattavudh Powdthavee (2007) found that half the British population view themselves as overweight, and that happiness and mental health are weaker among fatter people in Britain and Germany than among thinner folk. A partial explanation for this situation, say the authors, is that economic prosperity weakens self-control, encouraging some people to eat too much, which then undermines their subjective well-being. Taking a different tack Richard Layard (2005, Chapter 4) singles out our tendency toward social comparison with oneself and others as the basis for the unhappiness of many people concerning their own, comparable position of wealth. He concludes that 'one "secret to happiness" is to enjoy things as they are, without comparing them with anything better.' In other words, happiness cannot be found in trying to keep up with the Joneses. Another secret is to discover which things really make us happy (Layard, 2005, p. 53), what really counts in life.

Layard defines happiness as 'feeling good – enjoying life and wanting the feeling to be maintained' (p. 53). Elsewhere I have considered happiness under the headings of subjective and social well-being, two staples of the serious leisure perspective (Stebbins, 2009b, pp. 51–52). Unhappiness (lack of well-being), Layard says, is opposite of this, feeling bad and wishing things were different. Psychologist Martin Seligman (2003) brings us to the jumping off point for a positive sociology and the role of happiness therein when he argues that 'authentic happiness' comes from realizing our potential for enduring self-fulfillment. And, from what I have been saying about fulfillment in this book, it should be easy to conclude that, much of the time, no direct link exists between it and money.

The sociology of consumption further explains this observation. That is, can money buy either happiness, which is an emotion that expresses positiveness, or positiveness itself, which is an attitude (Stebbins, 2009b, p. 134)? Layard doubts it. Etzioni (2003, pp. 14–16) provides a review of some of the psychological literature supporting Layard's conviction. But, by stating the matter in the language of consumption, we may say that having money can help us more effectively pursue a fulfilling activity, for instance, with better skis, a finer violin, or additional cultural tourism. Money can make possible watching a film or dining out. In this formula money is *sometimes* a milestone along the road to (authentic) happiness. Meanwhile, as the section on nonconsumptive leisure demonstrated, many of the activities leading to positiveness and on to happiness do not hinge to any significant extent on this resource. In the final analysis both positiveness and happiness come from doing rather than having, even while having sufficient money occasionally facilitates the doing.

Authenticity and confidence

Only unusually, if at all, do authenticity and confidence find a true friend in money, in consumption. C.R. Snyder and Shane Lopez (2007, p. 241), who see authenticity as another facet of positive psychology, define it as: 'acknowledging and representing one's true self, values, beliefs, and behaviors to oneself and others.' Stated otherwise being authentic means being honest with oneself and with one's presentation of self to others. Charles Taylor (1991, Chapter 3) adds that being true to ourselves is a 'powerful moral ideal.' It is a form of personal sincerity. Still we must nuance Taylor's observation, by noting that our conception of authenticity is free of the connotation of moral correctness that is sometimes said to go with this trait. In other words, some people are quite capable of being authentically deviant, of freely admitting, for example, that they are members of an aberrant religious group, patronize a local nudist resort, or light up the occasional marijuana joint. Tolerable deviance (Stebbins, 1996c), of which these three are instances, though morally wrong in the larger community, is generally not so strongly stigmatized there as to force into inauthentic silence those who practice certain varieties of it.

People seem to find forced inauthenticity constraining and therefore disagreeable. By the same token they find they are attracted to situations where they 'can let their hair down' or 'be themselves'; these give a positive face to their existence. In this regard, Shane Silverberg (2008) learned, in a study of a Canadian energy company that one of several reasons its employees gave for liking the firm was that they could be themselves. No posturing required in their work setting, quite unlike some others they had experienced.

Authenticity is founded on, among other bases, one or more attractive personal and social identities. Having achievements in work or leisure, if not both, that the individual can be proud of and can therefore identify with, are themselves, real and genuine features of that person's sense of continuous, positive personal development as it unfolds over the years (Stebbins, 2009b, pp. 68, 93–94). It follows that most people would also like to see themselves in such terms (personal identity) and present themselves to, and be seen by, others in similar light (social identity).

Authenticity begets confidence through the process of *realistic* assessment of self and related achievements. In part this happens when, using skill, knowledge, and experience, we compare ourselves with others pursuing the same activity. A kind of informal, personal ranking of self and those others is thereby reached, as reflected in the social mirror into which each individual looks to see how he or she is viewed by those others. The authentic person accepts this reflected social assessment as reasonably accurate. Thus, a runner fond of participating in marathons who consistently places between the eightieth and ninetieth percentile of all contestants, if authentic, might say something on the order of 'I am a good marathoner,' but could not realistically say 'I am a champion marathoner' (i.e., consistently placing among the top four participants in each race entered). Given this record, other runners who know or know about this one would tend to describe this person in similar terms.

From these ideas on authenticity and confidence comes the conclusion that these personal orientations toward oneself cannot be bought in the marketplace. The good marathoner in the illustration above could not authentically claim that level of excellence in the race merely by purchasing and racing in the most expensive running shoes sold at the time. Facilitative consumption may help participants do better at their core leisure or devotee work activities. But, in the end, it is through their own efforts in doing those activities that authenticity and confidence emerge from, and from which positiveness and, ultimately, happiness eventually emerge as well. In this, the wonderful violin and engaging books discussed in earlier sections are no exception. One exception is the collector, for whom the purchased collectible rare stamp or painting, for example, is an authentic expression of that person's specialized knowledge of the area and, simultaneously, of his or her financial capacity to make the sale.

Conclusion

Much of what has been said in this chapter and the preceding one may be summarized in the following generalizations. One, consumption in relation to leisure, to the extent the first is either initiatory or facilitative of the second, is, in substantial part, a practical process – to be able to engage in the leisure, depending on its nature, the participant will have to buy a particular good or service. A second generalization follows, namely, that the heart of the consumption-based leisure experience, which is found in participating in the core activity or activities, lies outside this practical expenditure. This has been discussed in this chapter as Phase Two of consumption. Third, for some kinds of leisure, such monetary outlays are more or less unnecessary; this is the area of nonconsumptive leisure.

Nevertheless two exceptions to the first generalization should be underscored. One of them has already been treated of earlier in this chapter, namely, the special positive meaning some purchases come to have in association with the core serious leisure activity. The other is the role of conspicuous consumption in leisure (Veblen, 1899). Conspicuous consumption elevates significantly the importance for the consumer of the commercial side of this person's leisure. Purchasing expensive, dazzling goods and services earns the buyer special cachet in the admiring eyes of his or her contemporaries. To quote yet again Veblen's pithy statement: 'conspicuous consumption of valuable goods is a means of reputability to the gentleman of leisure' (1899, p. 64). Put in terms of our two-phase model, such commercial activity is a way wealthy people sometimes engage in initiatory consumption.

In sum, it seems that, whereas economists view the act of purchasing a good or service as lying at the heart of consumption, a leisure studies-based understanding of the consumptive process places the accent elsewhere. The latter stresses the first and second phases of consumption, and in doing this, minimizes the significance of the demarcating act of buying or renting something. Agreed, purchases relate in major ways to the economy, as manifested in wages earned, businesses sustained, and taxes collected, to mention a few indicators. But the motivational conditions and sociocultural contexts behind these ways are lost in this kind of analysis. In this respect a leisure studies perspective sheds important new light on modern-day acquisition and consumption.

Indeed it is by way of this perspective that the three issues covered near the end of this chapter have come to our attention. More particularly it was by discussing the ins and outs of Phase Two that the importance for consumption of uncontrollability and selfishness in leisure; consumption, happiness, and positiveness; and authenticity and confidence was brought to light. I have argued elsewhere (Stebbins, 2007d) that leisure studies is the world's only 'happy science'; it far more than any other discipline in the social sciences deals with the positive side of life. When looking through the lens of this field, one is thus predisposed to see that consumption – especially in Phase Two – is influenced by uncontrollability, selfishness, happiness, authenticity, and confidence.

In the next and final chapter the context within which leisure and consumption take place is once again the center of attention. There we shall see how social organization influences consumptive leisure and, simultaneously, may lead to social change.

6
Organizing for Consumptive Leisure

At various points throughout this book we have had a chance to reflect on one aspect or another of the vast contextual background against which consumption in the West occurs. Chapter 3 was the main locus of this interest. There we examined many of the personal and social conditions underlying the modern consumptive leisure experience. Conspicuous consumption was contextualized in this same manner in Chapter 2, while in Chapter 4 we looked briefly at the history of shopping and the relationship of gender and shopping, among other relevant social conditions. In fact, the theoretic approach of contextualizing leisure and consumption adopted in this book is evident on many of the pages of each chapter.

But for all that the job is not finished. There is still more to say on context, with some of it being covered in this chapter. In particular we are, to this point, missing a systematic discussion of the organizational basis of leisure and consumption. Brief mention in Chapter 2 of group and tribal identities hinted at this analytic slant. The present chapter begins with a statement of the conceptual framework to be used in exploring how organization influences and is influenced by consumption for leisure, as this occurs in different types of leisure organizations. The collectivities covered are small groups, social networks, and grassroots organizations as well as larger complex organizations and still more broadly, tribes, social worlds, and social movements (all discussed in detail in Stebbins, 2002).

Consumption in leisure collectivities

In the language of sociological theory we are actually speaking here of a main aspect of social, or institutional, organization. Guy Rocher defines the second as 'the total arrangement of all the elements which serve to structure social action into a whole, which has an image or a particular form which is different from its constituent parts and also different from other possible arrangements' (Rocher, 1972, p. 149). Interest in this chapter revolves around the collectivity, or the organizational, aspect of institutional organization. In leisure, as in many other areas of life, action is structured, or organized, in collectivities, notably, small groups, social networks, and grassroots organizations as well as in larger complex organizations and still more broadly, in tribes, social worlds, and social movements. Each structures the social behavior of its members in particular ways, some of those ways being unique to that kind of organization.

Still, Rocher's definition fails to consider a central aspect of organizational life: individual interests also structure organizations of every sort, including instituting them in the first place. Thus we may say about leisure organizations, as with all other kinds, that participation in them amounts to a two-way street of influence: individual to collectivity and collectivity to individual. Here the positive role of human agency crops up once again. This is the first of three critical assumptions on which this section is based.

The second critical assumption is that members of these different sorts of organizations know they are members. Third, these people value highly their membership, for given that we are considering only leisure organizations – those entered without coercion – members would abandon them were they truly disvalued. But when they are highly valued, belonging becomes an important motive for engaging in one or more of the leisure activities the organization promotes and facilitates.

Before going further into the matter of the organization of consumption and leisure, be aware that many casual, serious, and project-based leisure activities also allow for, if not require, solitary participation. They fall outside the scope of this section. Thus, someone may, in solitude, play the piano or the guitar, collect rocks or seashells, sit and daydream, or assemble from a kit a complicated electronic device. Volunteering is the principal exception to this rule. It is inherently organizational in the broad sense of the word, for by definition, it involves directly or indirectly serving other people, be they individuals or groups. What is meant by 'the broad sense' of the concept of organization?

'Organization' is used here as shorthand for the range of collectivities mentioned earlier in this chapter (dyads to social movements), entities that add social and psychological structure to leisure life. Accordingly discussion throughout will center primarily on these different collectivities manifested as leisure organizations rather than on the more abstract community or institutional organization of leisure, as seen in the sweeping communal arrangements that make available leisure services and opportunities. Additionally, since a fuller treatment is available elsewhere (Stebbins, 2002), we need to give in the present book only an *aperçu* of the different kinds of organization common in leisure. Finally, be aware that, while some of these organizations help explain consumptive leisure, they may also help explain consumption undertaken in the name of a work and nonwork obligation (e.g., a woman recommends to her daughter-in-law what the first holds to be the best vacuum cleaner or dishwasher soap on the market).

Small groups

This section examines leisure organization as manifested in dyads, or interpersonal relationships, triads, and other small groups of somewhat larger size. In sociological parlance dyads and triads are special variants of the small group. The latter is defined as a collectivity.

> small enough for all members to interact simultaneously, to talk to each other or at least to be known to each other. Another requirement is a minimum conviction of belonging to the group, a distinction between 'us,' the members of the group, and 'them,' the non-members.
>
> (Back, 1981, p. 320)

Moreover, dyads and other small groups endure, although only rarely for the lifetime of their members. At the same time, they are anything but evanescent. A gathering of passersby on a street corner animatedly discussing an automobile accident or two airplane passengers gabbing the whole flight through, then going their separate ways upon disembarking, fail to constitute a small group.

Small groups, whatever their size, generate their own 'idiocultures,' distinctive sets of shared ideas that emerge with reference to them (Fine, 1979). Idioculture is local culture, developed within and as an expression of an actual small group. It consists of a system of knowledge, beliefs, behaviors, and customs peculiar to that collectivity. Members use this system when interacting with one another, expecting they will be understood by other members.

Leisure is often enjoyed in dyads, exemplified in two friends organizing a surprise birthday party for someone or going together to the cinema. The triad is also a recognizable arrangement within which to partake of leisure, such as three people on a fishing trip or a classical music trio. And it is the same with larger small groups (e.g., church basketball team, several friends who hang out at a local jazz club, four couples who dine monthly at a restaurant). These three subtypes of the small group organization are found in all three forms of leisure, as the preceding parenthetic examples demonstrate.

Although the terms dyad and interpersonal relationship both refer to the two-person group, they emphasize different facets of it. The first points to numerical composition, whereas the second draws attention to the substantial level of intimacy and frequency of interaction existing between two people. Especially appealing in leisure dyads is the interpersonal component. Here each person through participating with the other in a given activity or set of activities gains a high level of deeply satisfying activity-related intimacy and interaction. This explains Kelly and Godbey's (1992, pp. 214–216) preference for describing leisure in small groups as 'relational leisure.' In this respect triads and slightly larger small groups are much the same.

In all these collectivities the interpersonal quality of the relationship helps explain the participants' motivation to engage in the leisure they share. It is not only that tennis is an interesting and challenging game or that sessions at the bar sometimes bubble with intrigue and gossip. It is also that these highly attractive activities are undertaken in pairs, triads, and small groups of people who are close friends, spouses, or partners, where the personalities of each hold mutual magnetic appeal based on such emotions and orientations as love, trust, respect, and affection.

How are these free-time small groups (dyads and triads included) related to consumption? One, in the intimate relationships on which these entities are based, members learn directly from the respected others in them about certain goods and services to buy, and others to avoid. Friends, partners, close relatives, and the like – I will treat of them summarily as 'intimates' – are usually people who can be trusted to avoid deceiving ego (the individual in question) and, to the best of their ability, give that person helpful advice. If a member of an informal small group of cyclists swears by the competence of a local repair service, the others may be inclined to patronize it. Or consider the man who accepts a recommendation from his wife to eat lunch in a center-city restaurant which she regularly frequents. He needs a reputable place to dine in with colleagues from out of town who are visiting his office, which is not, however, located in the center.

Two, it is plausible that the early adopters of new products and services mentioned in Chapter 2 are, compared with less familiar people, especially influential with their intimates in persuading them to buy the newly adopted good or service. We noted there that early adopters may be seen as authorities of sorts. And it is their intimates who know best these authorities and who are therefore inclined to trust their judgment and the information they give out about the good or service.

Three, people, when not shopping alone, are probably shopping with intimates. And each has a better idea than strangers or acquaintances normally would of what would appeal to the other. Therefore they can, as they shop together, point out attractive merchandise to each other, possibly even recommending its purchase. Buying a gift for an intimate is but a simple extension of this advisory arrangement. The intimate is better placed than nonintimates to know what the other person would like to receive.[1] All this applies to adults and possibly older adolescents, since young children would have, at best, a truncated idea of what, say, a parent would like. Conversely parents would typically be well versed on what their children would like as they browse together in a toy shop.

True, some of the purchases motivated by intimates are not related to leisure, but are instead related to a work or nonwork obligation. Be that as it may small groups play a major role in mediating leisure purchases, and we should add, rentals, that lead to acquisition of goods and services. Moreover such purchases motivated by intimates can augment the positive feelings bonding the people involved.

Four, intimates may also inform each other of relevant events and bargains on goods and services bearing on each other's leisure, precisely because they are sufficiently close to that person to know what he or she cares about in this area of life. So, Jane missed the advertised sale of runners in the morning's newspaper, but her husband saw it. He mentions it that evening, aware of her interest in replacing her current pair which, with use, has become seriously worn. Members of the jazz trio in which I play keep each other abreast of available workshops, concerts, and nightclub appearances bearing on our music. We also talk about amplification equipment, stand lights, new CDs, jazz festivals, music books, transcription programs, and so forth, all sold through local retailers or on the Internet. This social world of jazz music, like the vast majority of social worlds surrounding every serious leisure activity, is far too complex for one person to be constantly in the know about all of it.

And on a reduced scale of complexity, the same happens in certain types of casual leisure, most commonly in entertainment. Intimates point out upcoming concerts, new CDs, music festivals, books and magazines to each other, in case they have somehow missed this information. In short, intimates often serve an important informational function, and one often having consumptive consequences.

Social networks

Turning to social network the definition that best fits the small amount of work done on this form of organization within the domain of leisure is Elizabeth Bott's (1957, p. 59). Hers is simple: a *social network* is 'a set of social relationships for which there is no common boundary.' In the strict sense of the word, a network is not a structure, since it has no shared boundaries (boundaries recognized by everyone in the structure), no commonly recognized hierarchy, and no central coordinating agency. Nevertheless, links exist between others in the network, such that some members are directly in touch with each other while other members are not. For Bott the set of social relationships constituting a network is traced outward from a single person, or ego.

As individuals pursue their leisure interests, they commonly develop networks of friends and acquaintances related in one way or another to these interests. And, when a person acquires more such interests, the number of networks tends to grow accordingly, bearing in mind, however, that members of some of these will sometimes overlap. For instance, a few people in John's dog breeding network – they might be suppliers, veterinarians, or other breeders – are also part of his golf network – who might be suppliers, course personnel, or other golfers. Knowing about people's leisure networks helps explain how they, through positive agency, socially organize their free time. In this manner, as Tony Blackshaw and Jonathan Long (1998, p. 246) pointed out, we learn something new about leisure lifestyle and, we may now add, consumptive lifestyle.

Network ties can become conduits over which members hear about new goods and services concerning the leisure activity on which the network is based. To the extent the network is composed of close friends, this is the same kind of referring that occurs in small groups. Acquaintances in the network, however, may lack the authority of intimates, though this does depend on the status of the acquaintance. All collective amateur and hobbyist activities, in the West at least, are organized in part according to one or more local networks. Some members of these entities are more expert and committed to the activity in question than others, distinguished in Chapter 1 as core devotees, moderate devotees, and participants. Core devotees, though only acquaintances for some in the network, are nonetheless opinion leaders when it comes to, for instance, buying equipment or paying for services. Wayne Booth, an amateur cellist who was having trouble with 'scratchy' sound, provides a nice example:

> Friday night when we played cello quintets with an even better cellist, Gunther, and heard him praising his own cello, I asked him to try mine.
> He played a few riffs. It sounded to me wonderful. 'It's a much better cello than mine,' he said, 'and your bow is better than my bow.'
> 'So all the trouble is mine, not the cello's?'
> 'Well, it is a bit scratchy on the G-string. I'd try a different string.'
> Feeling like someone who has guiltily contemplated an affair with another lover, I welcomed 'her' back into my arms and today will take 'it' to Carl Becker [maker of Booth's instrument], for adjustment and maybe new strings.
>
> (Booth, 1999, p. 161)

Gunther, in this extract, was an acquaintance of Booth's, being in the eyes of the latter, something of a core devotee in chamber music.

From the standpoint of consumption, networks have at least one clear advantage over small groups. Networks are often larger – those on the Internet like Facebook, MySpace, and YouTube are huge – and as a result, bring ego into contact with more people (always acquaintances) as potential authorities on relevant goods and services. As always in both small groups and social networks, this authority is personalized in some convincing manner, such that recommendations are infused with trust in and respect for all parties involved. This atmosphere contrasts with that of the impersonal marketplace where *caveat emptor* is the watch-word and buyers, at least sophisticated ones, neither fully trust nor fully respect the recommendations of the sellers there.

Grassroots associations

At the next level of organization – the grassroots association – serious leisure predominates, even while some manifestations of it can also be found in casual leisure. The very, short-term nature of project-based leisure would seem to preclude such groups from developing around this form. According to David Smith (2000, p. 8).

> grassroots associations are locally based, significantly autonomous, volunteer-run formal nonprofit (i.e., voluntary) groups that manifest substantial voluntary altruism as groups and use the associational

form of organization and, thus, have official memberships of volunteers who perform most, and often all, of the work/activity done in and by these nonprofits.

The term 'formal' in this definition refers in fact to a scale of structure and operations that, in an actual association, may be informal, semiformal, or formal. Moreover, the line separating grassroots associations from paid-staff voluntary groups – treated of in the next section as volunteer organizations – is unavoidably fuzzy, distinguishing the two being primarily a matter of gradation. Both types fall under the heading of voluntary groups: 'nonprofit groups of any type, whether grassroots associations or based on paid staff, and whether local, national, or international in scope' (Smith, 2000, p. ix). According to this statement the following may be considered grassroots associations: Girl Guide troops, stamp collectors' societies, singles' clubs, outlaw biker gangs, and university fraternities and sororities.

With grassroots associations come new possibilities for disseminating information about goods and services. Through websites, newsletters, and in the larger organizations, magazines, members are exposed to discussions, advertisements, recommendations, and possibly even block and classified sales advertisements about relevant merchandise and services. Furthermore respected figures in the field may disseminate their own thoughts, endorsements, and recommendations. These lack the personal feel of their equivalents in the small groups and social networks, but they can nevertheless be authoritative.[2] Moreover participants in these associations, were they not members, might possibly miss such information.

Of course many members of grassroots associations are intimates of some other members. In this arrangement all the consumer-related information and its avenues of dissemination discussed in the sections on small groups and social networks operate here as well. Still many such associations are too large for all members to be so closely in touch with each other. Hence, from a consumer angle, this type of social organization adds a unique explanatory feature to our social organizational explanation of consumptive behavior. The same is true of volunteer organizations.

Volunteer organizations

Unlike the foregoing collectivities volunteer organizations offer leisure only to career and casual volunteers and to volunteers serving on projects. Volunteer organizations are distinguished by their reliance on paid staff, and by the fact that they are established to facilitate work for a cause or provision of a service rather than pursuit of a pastime. They nonetheless depend significantly on volunteer help to reach their objectives.

Pearce (1993, p. 15) holds that by far the largest number of volunteers work in these organizations. Yet some volunteer organizations may be staffed entirely by remunerated employees, volunteers only being engaged as unpaid members of their boards of directors. Hospitals and universities present two main examples. Many foundations can be similarly classified. Other volunteer organizations have a more even mix of paid and volunteer personnel; they include Greenpeace, Amnesty International, and the International Red Cross. Finally, some have only an employee or two, with all other work being carried out by volunteers. They are, at bottom, grassroots associations whose operations have grown complicated enough to justify paying someone to carry out those its volunteers are unable or unwilling to carry out.

With regard to the explanatory power of the organizational context of consumptive leisure, this type of collectivity is arguably is the weakest of those considered in this chapter. Why, because volunteers, compared with amateurs and hobbyists, tend to spend the least money on their leisure, whichever its form. Transportation-related expenses constitute a main cost, but in the typical case, they amount to rather little. Yet, volunteers might learn from each other, whether as intimates or acquaintances, about the most efficient way to reach the volunteer site and, if done by private car, the cheapest, safest, most convenient places to park nearby. A similar subculture of advice could emerge with reference to the clothing needed to carry out effectively and comfortably a volunteer role. Volunteers offer a service of some sort, and most of the time, they are its primary source. To do this, they may use a computer, microphone, telephone, automobile, or other implement, which however, is ordinarily supplied by the volunteer organization.

Leisure service organizations

Leisure service organizations are not voluntary groups, as just defined. Rather, they are collectivities run by a paid staff who provide one of more leisure services to a targeted clientele. To be sure, the clients are engaging in particular leisure activities, but the organizations providing them have goals that differ from the other leisure organizations covered in this chapter. Leisure service organizations are established either to make a profit, the goal of many a health spa, amusement park, and bowling center, for example, or to make just enough money to be able to offer their services. This is the goal of charitable, nation-wide, non-profit groups like the Boy Scouts and Girl Guides, the YMCA and YWCA, and the Elderhostel Programs as well as some governmental leisure and recreational programs and services.

Of concern in a book on leisure and consumption is the fact that many people in the West spend notable sums of money patronizing leisure service organizations. Intimates in small groups, social networks, and grassroots associations may, in ways already discussed, serve each other as pipelines of information and recommendation about some of these organizations. Where such organizations have a unique place in the realm of leisure and consumption, however, is through their goal of directly selling a service (and possibly some related products) to people seeking certain kinds of serious leisure experiences.

Paying for a leisure service can be relatively cheap or enormously expensive. Elite fitness clubs exemplify the latter, while municipal fitness centers number among the former. Nowadays, it appears, everyone must pay something for such services, whether public or private. Some municipal outdoor recreational facilities may be available free of charge (e.g., basketball courts, hockey rinks, tennis courts), but those indoors such as swimming pools, badminton courts, and gymnastic equipment generally ask a fee from users. In this respect the latter differ little from the fee-charging clubs and centers.

Before we leave this discussion of leisure service organizations, it is important to note that by no means all patrons go with leisure in mind to those that promote fitness. For some proportion of the adult population staying fit is a disagreeable obligation; were it somehow possible to be fit without the 'labor' of fitness training, they would jump at the alternative. Short of entering into a disquisition about why people engage in disagreeable activities (for more discussion on this matter, see Stebbins, 2009b, pp. 24–26), suffice it to say here that we must be careful in our analyses to distinguish patrons of leisure service organizations who are there for leisure from those who are not.

Social worlds

In Chapter 1, following Unruh (1980), I defined the concept of social world as a unit of social organization which is diffuse and amorphous in character. Generally larger than groups or organizations, social worlds are not necessarily defined by formal boundaries, membership lists, or spatial territory. A social world must be seen as an internally recognizable constellation of actors, organizations, events, and practices which have coalesced into a perceived sphere of interest and involvement for participants. Characteristically, a social world lacks a powerful centralized authority structure and is delimited by effective communication rather than by territory or formal group membership. We may add to this the observation that the richest development of social worlds may be observed in serious leisure, and if found at all in casual and project-based leisure, they are, by comparison, much simpler in composition.

Whichever the form of leisure the social worlds emerging around a given activity are largely impersonal. True, small groups (dyad and triads included) and social networks are found in every leisure social world, but intimate relationships of this kind are only one, typically small, component of this larger entity. Correspondingly, in addition to intimate relationships, consumption is also encouraged by way of various impersonal arrangements. The print and electronic media make up one class of such arrangements. Social worlds are served, in part, by newsletters, magazines, bulletins, and the like, all of which may be printed on paper or made available online, if not both. These media sometimes carry details about goods and services of interest to members of the social world. As already noted the media are typically the property of a grassroots association, volunteer organization, or leisure service organization.

Where social worlds add a unique layer to the explanation of consumptive behavior during the pursuit of leisure is through the four different types of members. These were described in Chapter 1 as strangers, tourists, regulars, and insiders (Unruh, 1979, 1980). The strangers are intermediaries who normally participate little in the leisure activity itself, but who nonetheless do something important to make it possible. These people may provide a service that must be bought, as repairmen do for musical instruments and some sports equipment, or goods that are needed to engage in a leisure activity (e.g., book stores – liberal arts hobbyists, kitchen shops – hobbyist cooks, magic supply stores – entertainment magicians). These are examples of serious leisure social worlds.

Still, some intermediaries sell not to regulars and insiders, as in the foregoing examples, but to tourists: the temporary participants in a social world who have come on the scene momentarily for entertainment, diversion, or profit. Thus, in the social world of rock music, agencies (strangers) sell tickets to rock music fans (tourists), who are seeking the enjoyment offered by a rock concert (casual leisure). At the concert sales of programs, refreshments, parking arrangements, and so on also fall into this category of consumptive leisure found through commercial contacts with strangers. Bars and restaurants renowned as hang outs for the fans of local professional sports teams are also strangers, who in this instance, sell food, drink, and possibly certain services before, during, and after games.[3] And museums, which typically charge an admission fee, also present visitors with opportunities to spend more of their money at the museum restaurant or its gift shop.

Should research ever be done on the matter, it would probably be found that substantially more cash is spent on casual leisure consumption of the sort just described – monetary exchanges between tourists and strangers – than on the serious leisure consumption of regulars and insiders when they patronize strangers. This hypothesis should hold, if for no other reason, than that many more people go in for casual leisure than for its serious relative. This is not to argue that everyone in town goes to museums, popular music concerts, professional sports matches and allied get-togethers. Rather it is that much larger proportions of the community go in for this sort of activity than pursue, as regulars or insiders, amateur art, science, sport, or entertainment.[4]

Tribes

Maffesoli's (1996) concept of the global, postmodern tribe was introduced in Chapter 2. Unlike the narrower, kinship-based tribes historically of considerable interest to anthropologists, his metaphorical conception refers to fragmented groupings left over from the preceding era of rampant mass consumption, groupings recognized today by their unique tastes, lifestyles, and form of social organization. Many of today's tribes exemplify well the group consumer identity, in interest of which members often buy goods and services to prove their status as belonging to one of them. Consequently, depending on the tribe, considerable money is spent on distinctive clothing, jewelry, and accessories as well as on music (e.g., instruments, compact discs, concert tickets, night club entrance fees).

Members of tribes ordinarily have some intimate relationships within this kind of collectivity. But, as global entities, the vast majority of members are only impersonally connected, chiefly by dint of their shared allegiance to the tribe and its *raison d'être*. It was observed in Chapter 2 that much of postmodern tribalization has taken place in the sphere of leisure, where it has given birth to a small number of interest-based, serious leisure tribes and a considerably larger number of taste-based, casual leisure tribes. The latter, which are especially popular among modern youth, are much less organized than the former, which also appeal to a much wider range of ages. That is in activity-based tribes composed of buffs (fans and buffs are compared in Chapter 2) a small number of leisure organizations provide their members with socially visible rallying points for individualized leisure identities as well as outlets for the central life interest they share (Stebbins, 2002, Chapter 5). Most often these organizations are clubs, which nevertheless serve as an important axis for the lifestyle enjoyed by enthusiasts pursuing the associated serious leisure activity.

The interrelationship between tribes, social worlds, casual leisure, and serious leisure along the dimension of complexity is presented in Figure 6.1.

At least three implications for leisure and consumption flow from the tribal form of organization. One, youthful members tend to buy identity symbols of the sort mentioned above as well as to spend money on the music they are attracted to. Musical concerts featuring favorite artists can easily be their greatest expense. Two, these youth are not commonly organized in clubs, as it typically happens for enthusiasts in the activity-based tribes. As a result those in the second are open to influence from the usual consumptive forces operating within grassroots associations.

Three, some people join in several tribes and, when tired of one of them, quit to become involved at their will in one or more others. The result nowadays for many participants, as Maffesoli (1996, p. 89) observed, is an individualized existence consisting of a multiplicity of lifestyles, of distinctive shared patterns of behavior organized around one or a handful of powerful interests. The central interest is the activity or cultural item sought: hairdo, clothing, bodily modification, style of music, and the various social practices associated with it. But when people add another tribe to their set of lifestyles, this usually requires

Taste-based tribes (e.g., music, clothing)	Activity-based tribes: fans (e.g., jazz, basketball)	Activity-based tribes: buffs (e.g., StarTrek, soap opera)	Social Worlds of (e.g., amateurs, career volunteers)
LEAST			MOST
COMPLEX			COMPLEX
Casual leisure	Casual leisure	Serious leisure	Serious leisure

Figure 6.1 Structural Complexity: From Tribes to Social Worlds
Source: From R.A. Stebbins, *The Organizational Basis of Leisure Participation: A Motivational Exploration*. State College, PA: Venture, p. 70.

new expenditures. The newcomer must purchase the clothing, hair style, accessories, and so on that mark members of that tribe. In the end the main motive for participation is to join in the 'warm and fuzzy' company of kindred spirits to hear the tribe's music or parade oneself dressed in its characteristic garb.

Social movements

What remains to be examined in this long section on leisure and organization is the social movement. A *social movement* is a noninstitutionalized set of networks, small groups, and formal organizations that has coalesced around a significant value, which inspires members to promote or resist change with reference to it. The first question is whether participation in a social movement is a leisure activity. The answer is both yes and no, for it depends on the movement in question. Movements abound that gain members by their own volition, suggesting that the members experience no significant coercion to become involved. Some religious movements serve as examples, as do movements centered on values like physical fitness and healthy eating. Still, the latter two also include people who feel pressured by outside forces to participate, as when their physician prescribes exercise and weight loss or face likelihood of an early death. In such cases negative emotions such as fear, hate, disgust, and revenge dominate, pushing the troubled person into direct movement participation. Thus some social movements are composed of enthusiasts who are there for leisure reasons and other people who are compelled to be there (disagreeable obligation).

Finally, there are movements that seem to find their impetus primarily in people who feel driven to champion a particular cause, such as the celebrated temperance movement of early last century and the vigorous antismoking movement of modern times. A strong sense of obligation also fuels participation in them. Those who make up the gun control and nuclear disarmament movements seem cut from this cloth as well. Whether such people are taking leisure must be determined empirically through direct research on their motivation to participate.

Social movements are not social groups, whatever their size, but rather networks of groups, organizations, and individuals spanning a community, region, society, and these days with rampant globalization, even much of the world. For individual participants, intimate interaction is therefore possible with only a very small proportion of all members of the movement. Moreover a cardinal feature of any social movement is its ideology, one function of which is to justify and motivate pursuit of the key values shared by the participants. On the most general plane, these values are initiation or prevention of what they see as important social changes.

Ralph Turner and Lewis Killian (1987, p. 223) hold that a social movement is, at bottom, a collectivity; it is 'something of an interrelated and coacting unity of persons, rather than a mere aggregate of persons acting separately but in parallel fashion.' As such, a social movement is more than a tendency or a trend to, for example, buy certain brands of cars or watch certain types of television programs, something usually done by aggregates of people acting separately but in parallel. Furthermore movements are comprised of acting individuals, which excludes those who accept movement values but do nothing to help realize them. As such this is a sociological rather than a psychological question; the essence of collective behavior, of which the social movement is a main type, is found in the actions of sets of people, not in the acts of isolated individuals.

Some social movements, be their members primarily of the leisure variety, the forced variety, or a combination of the two, have left an indelible mark on modern and postmodern life. Thomas Homer-Dixon (2007), for instance, commenting on the success of the mothers' movement in the 1960s, which championed banning atmospheric nuclear testing, a practice that contaminated children's milk, now urges a similar formation fired by the goal of trying to bring global warming under control. Considered alone a social movement is a distinctive form of organization, which among other things, provides serious and casual leisure for volunteers. Further it may possibly provide leisure projects for volunteers, enabling the latter to become involved with a movement for a limited amount of time. Examples include participating in a fundraising campaign, organizing a major rally, and lobbying for a crucial piece of legislation.

What unique opportunities for consumptive leisure are made possible by freely participating in a social movement? In general, the answer to this question appears to be that there are no unique opportunities of this kind. For the most part the leisure activity engaged in under the aegis of a social movement is some genre of volunteering. Expenses here seem to be the same as those encountered while volunteering in other areas of the nonprofit sector. Further social movements encompass a myriad of dyads, triads, small groups, and social networks and, for those that endure for some time, grassroots associations and at least one volunteer organization. All the avenues of consumptive influence operating in these entities also operate when the entities are part of a movement. Incidentally committed members of social movements often donate money to the cause for which they are so passionately working. But donations of any kind fail to qualify as consumption. They also fail to qualify as leisure (Stebbins, 2007a, p. 10).

This said the social movement is the only type of organization discussed in this chapter where participants are motivated by strong commitment to a cause. Or, if members of a small group or volunteer organization as so motivated, it is likely these collectivities are embedded in a larger movement whose commitment they share. I know of no research into the effects such commitment has on consumption. Yet there is reason to believe that it could be considerable. It might well turn out, upon careful research, that highly committed people in social movements are inclined to splurge on movement-related goods and services, while being much less spendthrift (perhaps because there is now less money) on consumables than in other areas of their lives. For instance, committed members might spend disproportionately on pertinent items like books and magazines, admission fees and transportation costs to special events (e.g., talks, conferences, workshops), and relevant equipment and apparel (e.g., religious paraphernalia, special jacket). To the extent these committed members are enthusiasts and not coerced into the movement by strong negative sentiments, these purchases are also made in the name of leisure.

Consumer-related social movements

Today leisure and consumption are joined, among other ways, in a variety of overlapping social movements, two of which will be taken up here. By considering leisure and consumption in this expanded context – both movements are global – we may broaden further our understanding of the relationship of the two as this combination of human endeavor fits with social life in the 21st century. In particular this is field of action for some of the citizen – consumers discussed in Chapter 4. Consumer-related social movements vividly demonstrate the importance of the two-way influence between consumption and leisure organization.

The two movements to be examined are simple living and sustainability. Both, according to David Aberle's (1966) classification, are most accurately classified as 'reformative movements.' Reformative movements aim for partial change of the social order. Many of them work toward some sort of social reform, with the antiabortion and environmentalist movements being prime modern examples. The campaign for nuclear disarmament is reformative in nature. In fact, the list of contemporary reformative movements is long.

Aberle's four-fold classification of social movements is still considered one of the most useful and comprehensive. It is constructed along two dimensions: locus and amount of desired change. Concerning the first, some movements strive to change individuals while others strive to change the social order. The amount of change sought may be partial or total. Cross-classifying these two dimensions results in four types of movements labeled by Aberle as transformative (total or near total change in the social order), reformative (partial change of the social order), redemptive (total change in individuals), and alternative (partial change in people).

In all four types there is, at just pointed out, a variety of links between leisure and consumption within the conceptual framework of social movement. But this has so far been an individualistic approach to organization and consumption: how individuals are motivated to buy goods and services by dint of their membership in one or more collectivities. From here on our focus changes, however, as we look at how one type of organization – the social movement – is changing, or is at least attempting to change, the consumptive ways of society, whether that society is national, regional, or global.

The scope of the two movements covered below is far wider than consumption. Consequently discussion of them will be limited as much as possible to consumption and its relationship to leisure.

Simple living movement

In Chapter 2, we briefly considered voluntary simplicity and its distinctive relationship to conspicuous consumption. The time has now come to examine more generally the relationship of consumption and simplicity. The spirit of living simply energizes a growing social movement today, which promotes the special lifestyle of voluntary simplicity. In a book by this title Duane Elgin (1981), himself heavily influenced by Gandhi, writes:

> that, among other things, it is a way of living that accepts the responsibility for developing our human potentials, as well as for contributing to the well-being of the world of which we are an inseparable part; a paring back of the superficial aspects of our lives so as to allow more time and energy to develop the heartfelt aspects of our lives.

The simple living movement, – also know by such denominations as 'downshifting' and 'creative simplicity,' among others – was launched in the mid-1930s with an article written by Richard Gregg (see Elgin, 1981, pp. 297–298, for bibliographic information on the several reprinted versions of this article). Still, the two quotations below suggest that need for the movement is much older:

> Better is an handful with quietness, than both the hands full with travail and vexation of spirit.
>
> Ecclesiastes 4:6

> Half our life is spent trying to find something to do with the time we have rushed through life trying to save.
>
> Will Rogers, *Autobiography*[5]

As a practical strategy voluntary simplicity may be seen as cutting back on something held by a person to be unnecessary. True voluntary simplifiers – the people ideologically motivated by the movement to create a lifestyle based as fully as possible on the principles of voluntary simplicity – go much farther than a single practice, exemplified in driving a compact car instead of a sport utility vehicle or growing their own vegetables instead of buying them at the supermarket. Still voluntary simplicity may be pursued in degrees ranging from downsizing the family automobile or growing vegetables to a more completely self-sufficient existence consisting of, among other things, walking and using public transit, making one's own clothing, living in a home no larger than absolutely necessary, and resorting wherever possible to do-it-yourself to meet all domestic obligations. For the purposes of this book voluntary simplicity refers to this entire range of practices leading to a more or less simpler lifestyle than before.

As a lifestyle balance strategy the search for a new level of simplicity opens up the possibility of becoming less dependent on the paying job as a whole or on some of its key obligations. One might ask, 'Should I need to earn $100,000 annually, were I to drive a cheap, economical car or reduce the size of my house or apartment?' Or 'should I need such a job were I to perform my own yard work rather than meet this obligation by hiring a costly commercial service?' Voluntary simplicity enables its followers to live on reduced income, commonly achieved by, in some way, decreasing the amount of money they allot to managing their nonwork obligations.

We turn next to another implication of Elgin's conception of voluntary simplicity. It is that, in effecting a lifestyle truly consistent with the tenets of simple living, enthusiasts for this lifestyle also appear destined both to increase their list of nonwork obligations and to reduce the amount of free time in which 'heartfelt' leisure may be pursued. Consider that living simply may require a person to, for instance, walk and use public transit (in lieu of driving a car), take recyclable trash to a recycling depot (in lieu of sending it to the municipal landfill), grow vegetables or bake bread (in lieu of buying these items at a supermarket), and acquire and use wood for home heating (in lieu of purchasing gas or oil for this purpose). Some of these simple living obligations might well be seen by some enthusiasts as pleasant, as essentially leisure, including tending a garden, baking bread, and even chopping wood for home heating.

But all these activities take time, which has to be found in a person's weekly accumulation of free time. Moreover when the activity is disagreeable, this robbing of Peter to pay Paul cuts into the hours that might be used for self-fulfilling serious leisure. It also cuts into time for casual leisure, consequently weakening access to, or the experience of, the previously mentioned benefits it can offer. What is more this way of shifting to a simpler existence leaves fewer of these activities for rounding out an optimal leisure lifestyle. Finally people, to the extent they are occupied with both work and nonwork obligations, now have, when it comes to organizing their daily lives, significantly less room for maneuver.

It was noted in Chapter 2 that some voluntary simplifiers are categorically anticonsumerist; they try to eschew all purchases, save those meeting basic needs. Some are also 'non-adoptionist,' in that some non-adopters not only like certain retro goods but also want to economize on their purchases. What these tendencies mean for consumption remains to be systematically explored through research. That voluntary simplifiers resort wherever possible to do-it-yourself, may turn out to mean that, compared with ordinary consumers, they buy fewer services but more goods with which to carry out their own money-saving activities and projects.

The simple living movement is a sign that life in the West has grown much too complex for many people, and that it has been this way for some time. Gregg first broached the subject in modern times in the mid-1930s. He was influenced by the early 20th-century writings of Gandhi, however, while before that, in the first half of the 19th century, Henry David Thoreau wrote, in distinctive Thoreauvian style, about simplicity. Somewhat more recently, in 1959, Julie London popularized the song 'Give Me the Simple Life' and, in 1961 in London, the musical 'Stop the World, I want to get off' opened, leading eventually to a run of 555 shows. These authors and artists as well as the writers of the two preceding quotations show that, long before it became the popular movement it is today, simple living was on the minds of many an individual thinker.

One logical conclusion of simple living would have us minimize the so-called superficial, casual-leisure aspects of life. Still, as argued in Chapter 1 (see also Stebbins, 2007a, pp. 41–43), casual leisure has its important benefits, while also serving as an indispensable element in the personal formula for creating an optimal leisure lifestyle. In other words, to put in proper focus the movement's call for a reduction of life's superficialities, the serious leisure perspective suggests that the simple living movement should have as one of its primary goals personal development of an optimal leisure lifestyle.

A finely nuanced understanding of simple living, such as just suggested, is no trifle. Personal fulfillment and well-being are universal human values, which stand out, in part because so many people the world over fail to attain them. And now, with the wolf of global warming at the door, this movement becomes all the more important for the central role it can play in helping combat some of the causes of this portentous environmental condition. Put otherwise simple living is related, in a complicated way, to another contemporary movement, that of sustainable consumption.

Sustainable consumption

The Oslo Symposium on Sustainable Consumption, held in 1994, among its other accomplishments, proposed a definition of the idea:

> the use of services and related products which respond to basic needs and bring a better quality of life while minimising the use of natural resources and toxic materials as well as emissions of waste and pollutants over the life cycle of the service or product so as not to jeopardise the needs of future generations.
> (Norwegian Ministry of the Environment, 1994)

There is much in this definition that conforms to Elgin's definition of voluntary simplicity. Yet there are important differences, too, seen in the additional emphasis in simple living on ways of life that help us realize our human potential and allow more time and energy for development of the heartfelt aspects of our lives. Phrased in the language of positive sociology, both movements have in common a negative agenda of solving environmental problems, but only the simple living movement has a second, equally powerful, positive agenda (on the distinction between negative and positive, see (Stebbins, 2009b). This positive agenda is twofold: to develop human potential – that is, engender fulfillment – and to develop the heartfelt aspect of life – that is, encourage people to engage in activities which make their existence attractive, worth living.

The social practices model championed by Gert Spaargaren (2005, p. 16) harmonizes well with the tenets of the simple living movement. He contrasts his model with the dominant social psychological attitude-behavior model. Thus the center of the social practices model is composed not of individual attitudes and norms but instead of 'actual behavioural practices, situated in time and space, which an individual shares with other human agents' (p. 16). Second, this (sociological) model calls attention to particular groups of people and their lifestyles and to the possibility that they may reduce the overall environmental impact of their everyday routines by engaging in enlightened social practices related to food, sport, travel, leisure, clothing, and shelter. Third, Spaargaren's model enables us to analyze the process of reducing the environmental impact of consumption in these and other critical domains of social life. In this regard we should study the actions of knowledgeable, capable consumers who take advantage of the commercial opportunities they find in particular sectors of the economy. Social structures are thereby pushed to the center of analysis, not relegated to the periphery, as happens in the attitude-behavior model.

This model fits also comfortably with some of the key concepts advanced in this book. Thus Spaargaren's 'actual behavioural practices, situated in time and space' are in tune with the concept of activity that has been one of our principal leitmotifs throughout. I interpret the 'particular groups of people' to be, in the area of leisure, a reference to the types and subtypes of amateurs, hobbyists, volunteers, and casual leisure participants when acting as purchasers of relevant goods and services. Third, his model directs attention to 'identifiable domains of life.' Here consumers (e.g., the amateurs, hobbyists, etc.) in their particular groups and through their lifestyles act as agents of change in harmful environmental practices. The lifestyles are the ones we have been referring to throughout as optimal leisure and simple living lifestyles.

Michael Maniates (2002) examines the contrasting popular and scholarly appreciations of voluntary simplicity and sustainable consumption:

Curiously, the media's fascination with simplicity has not been matched by attention in more scholarly circles. One reason, I explain shortly, is that voluntary simplicity butts up against mainstream environmentalism's understanding of 'sustainability' and 'the consumption problem' and is marginalized as a result. By its very existence, the VSM [voluntary simplicity movement] insists that real reductions in consumption – at least for some, framed in particular ways – bring real nets benefits to be enjoyed rather than sacrifices to be endured. But mainstream environmentalists – not to mention policymakers, planners, and many academics – find it difficult to entertain and give voice to the distinct possibility that doing with less could mean doing better and being happier. Locked in a calculus of sacrifice, activists and academics and policymakers alike tend first to romanticize the VSM, and then to marginalize it to the domain of 'fringe' activity that is oddly interesting but fundamentally irrelevant to the practical politics of suistainability.

(Maniates, 2002, pp. 201–202)

Using the terminology introduced earlier in this book, the sustainable consumption movement is largely couched in the negative orientation of problem-solving, whereas the movement that has sprung up around voluntary simplicity also has an unmistakable positive tone about it.

The problem-solving orientation of the sustainable consumption movement manifests itself in diverse ways. One, people are directed to adopt a lifestyle that reduces individual and communal use of the Earth's available natural resources. This should be accomplished by altering wasteful and environmentally harmful means of transportation and by changing patterns of energy consumption and types of diet (Winter, 2007). Two, success in this endeavor leads to sustainability and a natural balance with the Earth's ecological system and its natural cycles in what amounts to a symbiotic relationship between the two. Three, sustainable consumption and the philosophy of ecological living that lies behind it go hand in glove with the principles of sustainable development. The problem confronting this movement is to minimize the 'ecological footprints' left by mankind, where these incursions affect the environment unfavorably. The hope in all this is to preserve the planet for future generations of human beings and other life.

Leisure enters the discourse on sustainability by at least two routes: what people do to the environment during their free-time activities and what changes to the environment must be made to make these activities possible. Thus, golf courses and alpine ski hills are inimical to the principle of sustainability in the measure that, in preparing them, large numbers of trees must be removed. With this go the habitats of the resident animals and birds. Animals living outside the courses and hills may find their access to water disturbed as well. Further, in certain areas of the world, golf courses require continual watering, creating in some cases a serious drain on local supplies. Once the courses and hills are completed golfers and skiers pay money to use them, which they do with little further deleterious impact on the environment.

By contrast many beaches need no modification for swimming and sunning to be enjoyed there, though both, especially when done in large numbers, may change them for the worse. This they do mostly by leaving refuse and polluting the water. The same scenario may unfold with respect to back-country camping. Likewise some terrain cannot withstand the pressure of uncontrolled snowmobiling, at present a prickly issue in, for example, Yellowstone National Park in the United States.

A federal court ruled on Sept. 15, 2008 that the Bush Administration's decision authorizing snowmobile use in Yellowstone National Park violates the fundamental legal responsibility of the National Park Service to give top priority to conservation of national park resources. The court found that the Administration authorized snowmobile use despite scientific conclusions by the National Park Service that its decision would result in significant increases in disruptive noise, unhealthy exhaust and harm to Yellowstone's animals.

(http://news.greateryellowstone.org/node/154,
retrieved 11 January 2009)

The natural environment also includes the wildlife living in it. For example, frequently offering human food to wild animals and birds can result in an unintended modification of faunal nature. When fed this way these creatures can grow dependent on this food supply, with consequences such as gaining too much weight for effective survival in the true, nonhuman, wilderness (e.g., overweight chipmunks and squirrels in parks) and raiding human sources of food to satisfy their new appetite. True, bears, one the main raiders of human sources, are not usually intentionally fed, as a hiker might feed a bird or ground squirrel. Instead bears develop a taste for human food when carelessly but inadvertently left out for them to eat.

Look at the adverse effect hunting can have on large animal species. Robert Britt writes in *Live Science*, on online scientific newsletter, that 'survival of the smallest is not exactly what Darwin had in mind, but in some animals species, humans may be forcing a smaller-is-better scenario, and the ultimate outcome may be species demise.' He says the process runs as follows: 'Bigger males with bigger horns tend to father bigger offspring, causing the average size of a species to increase over time. With hunters targeting these trophies, smaller males are more successful at mating, so their genes are spread through the population more effectively and the average size shrinks' (http://www.livescience.com/animals/090106-reverse-evolution.html, retrieved 13 January 2009). This can hardly be regarded as leisure-based sustainable consumption. For hobbyist hunters, who pay for such goods and services as shells, rifles, transportation to hunting areas, and taxidermal work on trophies, are changing the natural evolutionary process of the species they hunt.

Sustainable tourism has, of late, been a much talked-about issue in the fields of tourism, scholarly research, and governmental policy. Expressed most simply, sustainable tourism can be said to be: 'tourism that takes full account of its current and future economic, social and environmental impacts, addressing the needs of visitors, the industry, the environment and host communities' (United Nations Environment Programme and World Tourism Organization, 2005, p. 12). Obviously tourism is quite capable of turning into unsustainable leisure activity. The huge numbers of people traveling around the globe these days for touristic purposes contribute to pollution of the air by jet liners. Another part of the problem is the 'carrying capacity' of the tourist sites. Immensely popular destinations such as Paris, New York, and Amsterdam may find their water, sanitation, and electrical services strained at times by the demands placed on them by the seemingly endless streams of visitors. Touristic pressure on the Galapagos Islands, located off the Ecuadorian coast, has forced authorities to restrict the number of tourists allowed to visit them each year. Excessive tourism is one reason why, in 2007, UNESCO placed the Galapagos on its 'World Heritage in Danger' List.

Finally it is instructive to consider the group of nature challenge activities discussed in Chapter 5 under the heading of nonconsumptive leisure. They attract both locals and tourists, many of whom, however, spend comparatively little money (especially the locals). For this reason they tend to fall outside the field of sustainable consumption. But from another angle they may be seen as directly related to it, in that sustaining nature in its pristine state is a major goal of all nature challenge participants. A cardinal reason for engaging in a particular nature challenge activity is to experience its core activities in a natural setting. In other words, while executing this activity, the special (aesthetic) appeal of the natural environment in which this process occurs simultaneously sets the challenge the participant seeks (Stebbins, 2009c).

A study of mountain kayakers, snowboarders, and alpinists explored their interest in pursuing the corresponding nature challenge activities in the awe-inspiring natural environment in which they occur (Stebbins, 2005b). Being in awe, or wonder, of an aesthetically attractive natural setting brings an unforgettable psychological dimension to the activity in question. For the kayakers their wonder-filled environment had, among other features, the sound, sight, and feel of the rushing mountain rivers and creeks and the rock, earth, trees, and vegetation through which they flow. The snowboarders were enthralled with the snow-filled back-country setting through which they rode as it descended, often precipitously, before them. The mountain climbers loved rock, its nooks, crannies, and solidity (when present) and the way these qualities and others combined to create a sense of being suspended in air far above the base of the steep slopes they mounted. These participants want these settings sustained largely as they are, and to this end, some of them militate through such organizations as the Sierra Club and American Trails against logging, road building, commercial development, and the like in the area.[6]

Conclusions

Our tour of the hybrid field of leisure and consumption, made possible by the six chapters of this book, has, at times, had a negative feel about it. This sentiment has occasionally clashed with the claim, made in greatest detail in Chapter 5, that leisure studies is the 'happy science.' No other social science can claim this title. For that science has as its central interest the core activities that bring (dare I say tempt or lure) participants to the larger, general leisure activity. We defined core activity as a set of interrelated actions or steps that must be followed to achieve an outcome or product that the participant in the larger leisure activity finds attractive. In broad terms, leisure studies examines the activities (core and general) pursued in free time, the people who engage in these positive activities, the processes of leisure provision, the policies that direct these processes, and the ramifications of these four for all levels of society, local, national, and international.

But consumption in the name of leisure, as has been shown here, sometimes places the second in negative light. Many of the deleterious effects of what has been dubbed 'over-consumption' stem from people pursuing free-time activity. In Chapter 2 we saw how some forms of conspicuous consumption and conspicuous waste can be traced to leisure. In Chapter 3, except in the writings of Nazareth and Douglas and Isherwood, it is consumption that the authors covered there most often excoriate for its pernicious effects on individual and society. In these condemnations leisure is rarely mentioned, and then, only generally. Chapters 4 and 5 are, we might say, the happiest of this book, in the sense that the consumptive behavior discussed there is often related in favorable ways to leisure interests. Still the discussion of wants and needs lays bare the occasionally negative features of leisure shopping, as does deviant consumption. Chapter 5 hums along mostly on a positive note, although the section on uncontrollability and selfishness introduces again some negativeness. As for the present chapter the section on sustainable consumption has the most negative sense about it.

The sweet and sour mix of positive and negative sentiments toward leisure brought to light when juxtaposed with the consumption carried out with reference to it constitutes still another reason for considering leisure and consumption together. Leisure is happy activity, and it is good to know that its pursuit can provide some needed positiveness in today's all-too-often depressingly negative world (Stebbins, 2009b). But when discussed with consumption we are forced to nuance our understanding of leisure as fully happy, fully positive. The warning that, unless carefully monitored, leisure may generate negative consequences for both individual and society is a principle we must constantly bear in mind.

Notes

1 The Nature of Leisure and Consumption

1. I argue elsewhere (Stebbins, 2009b) that, at the *activity* level, all of everyday life may be conceptualized as being experienced in one of three domains: work, leisure, and nonwork obligation.
2. I am aware that standard sociological theory conceives of roles as dynamic and statuses as static. Compared with activities, however, roles are *relatively* static.
3. The serious leisure perspective and its three forms are discussed in considerably greater detail in three consecutive reviews of it (Stebbins, 1992, 2001a, 2007a).
4. For papers on serial murder and violence done for 'fun,' see the special issue of *Leisure/Loisir* mentioned above.

2 Conspicuous Consumption

1. In some earlier publications I have occasionally referred to fans as consumers. In this book on consumption such confusing nomenclature will not do; fans and consumers will not be treated of synonymously.
2. The relationship of the three forms of leisure and simple living is examined in greater detail in Stebbins (2007c).

3 Consumption and Leisure in Context

1. See, for example, Clive Jenkins and Barrie Sherman (1981) and Barrie Sherman (1986). Rifkin comes closer than the others to completing the story when he examines the role of volunteers and community service in the closing chapters of his book.

4 Phase One: Shopping

1. Some purchases are made as a series of installment payments or as a debt (nowadays often by credit card) to be paid later. In the meantime the buyer, now in Phase Two of consumption, has immediate use of the acquired good or service.
2. For a more detailed account of the history of modern shopping in the West, see Harris (2005, pp. 237–243).
3. This paragraph and the next draw heavily on the fine summary of the history and geography of consumption presented by Clarke, Doel, and Housiaux (2003, pp. 17–20, 27).

4. Then or now, there has never been a female equivalent, a *flâneuse*. See Janet Wolff (1985, p. 44).

5. The popular understanding of obligation, which fails to recognize obligation in leisure, leads to confusion when the obligation in question is agreeable. It is therefore important to remember that meeting an unpleasant obligation is not, by definition, an act of volunteering. Here nothing is voluntary and, if it is not voluntary, it is not leisure (Stebbins, 2009b, pp. 24–26).

6. Of interest here is the fact that double-blind tests have demonstrated that local tap water is often identified by participants in experiments as tasting as good if not better than the commercially ballyhooed stuff in the bottle (Swartzberg, 2007, p. 3).

7. As mentioned earlier Campbell's (1997) data show the distaste of the disgruntled shopper-husband in the typical tourist couple; he 'hates shopping' as much as his wife loves it.

8. Moubarac, Gupta, and Martin (2007, p. 521) maintain that online poker is significantly less skilled, because players are unable to read each other's gestures and facial expressions.

5 Phase Two: Consuming the Purchase

1. True, the violin may have been borrowed, even stolen. Both alternatives are, however, beyond the scope of this book (see Chapter 1).

2. A nature challenge activity (NCA) is a type of outdoor pursuit whose core activity or activities center on meeting a natural test posed by one or more of these six elements. A main reason for engaging in a particular NCA is to experience participation in its core activities pursued in a natural setting. In other words, while executing these activities, the special (aesthetic) appeal of the natural environment in which this process occurs simultaneously sets the challenge the participant seeks (Stebbins, 2009c).

3. Other kinds of theater such as puppetry, clowning, cinematic production, and entertainment magic typically entail considerable expense.

4. One of my respondents in a study of amateur baseball (Stebbins, 1979) actually said as much, though he said he loved his girlfriend more than he loved football (which he did not play). He did not seem to be joking.

6 Organizing for Consumptive Leisure

1. It is common practice for people who want to buy a gift for someone at the club or office, for example, to ask one of the recipient's intimates about his or her tastes and interests in an attempt to give something truly attractive.

2. In fact the recommendation may be powerfully authoritative, in good part because the recommender is a distant member, though outstanding exemplar, of the field. Few golfers would claim Tiger Woods as an intimate, but many would take seriously his endorsement of a certain kind or brand of golf club. Is it not the same when respected professionals endorse, for example, a brand of guitar, amplifier, basketball shoe, or alpine ski?

3. These services might include transportation to and from the establishment to the site of the game. For fans unable or unwilling to attend games and

for games played in other communities, the bar or restaurant may offer televised coverage of these matches. With such services this class of stranger is attempting to turn a profit on the sale of food and drink to leisure tourists.

4. I estimate – and it is a crude calculation – that on average about 20 percent of the overall population is involved in serious leisure of some kind (Stebbins, 2007a, p. 134).

5. Rogers' autobiography was published posthumously in 1949 and consists, in good part, of pithy quotations made during his professional life, which began just after the Boer War.

6. A properly sustained setting would nevertheless include some minor modification of natural processes, for example, removing fallen trees that impede kayaking and hiking, reconstructing washouts on trails, and setting tracks in snow for cross-country skiing.

Bibliography

Aberle, D.F. (1966). *The Peyote Religion among the Navaho*. Chicago. IL: Aldine.

Arai, S.M., and Pedlar, A.M. (1997). Building communities through leisure: Citizen participation in a healthy communities initiative. *Journal of Leisure Research*, 29, 167–182.

Aronowitz, S., and DiFazio, W. (1994). *The Jobless Future: Sci-Tech and the Dogma of Work*. Minneapolis, MN: University of Minnesota Press.

Back, K.W. (1981). Small groups. In M. Rosenberg and R.H. Turner (Eds), *Social Psychology* (pp. 320–343). New York: Basic Books.

Barber, B.R. (2007). *Consumed: How Markets Corrupt Children, Infantilize Adults, and Swallow Citizens Whole*. New York: W.W. Norton.

Baudrillard, J. (1981). *For a Critique of the Political Economy of the Sign*, trans. C. Levin. New York: Telos Press.

Baudrillard, J. (1998). *The Consumer Society: Myths and Structures*, trans. C. Turner. Thousand Oaks, CA: Sage.

Bauman, Z. (2003). Industrialism, consumerism and power. In D.B. Clarke, M.A. Doel, and K.M.L. Housiaux (Eds), *The Consumption Reader* (pp. 54–61). London: Routledge.

Belk, R. (2007). Consumption, mass consumption, and consumer culture. In G. Ritzer (Ed.), *The Blackwell Encyclopedia of the Social Sciences* (pp. 737–746). Cambridge, MA: Blackwell.

Bell, D. (1970). The cultural contradictions of capitalism. *The Public Interest*, 21 (Fall), 16–43.

Benjamin, W. (1973). *Charles Baudelaire: A Lyric Poet in the Era of High Capitalism*, trans. H. Zohn. London: NLB.

Bennett, A. (2003). Subcultures or neo-tribes? Rethinking the relationship between youth, style, and musical taste. In D.B. Clarke, M.A. Doel, and K.M.L. Housiaux (Eds), *The Consumption Reader* (pp. 152–156). London: Routledge.

Binham, C. (2008). Yes, guv'ner, U.K. needs butlers. *Calgary Herald*, Sunday, 13 April, p. D1.

Blackshaw, T., and Long, J. (1998). A critical examination of the advantages of investigating community and leisure from a social network perspective. *Leisure Studies*, 17, 233–248.

Booth, W. (1999). *For the Love of It: Amateuring and Its Rivals*. Chicago, IL: University of Chicago Press.

Bott, E. (1957). *Family and Social Network*. London, UK: Tavistock Publications.

Bourdieu, P. (1977). *Outline of a Theory of Practice*. Cambridge & New York: Cambridge University Press.

Bourdieu, P. (1979). *La Distinction: Critique Sociale du Jugement*, trans. R. Nice. Paris: Les Editions de Minuit.

Bowen, C.D. (1935). *Friends and Fiddlers*. Boston, MA: Little Brown.

Bowlby, R. (1997). Supermarket futures. In P. Falk and C. Campbell (Eds), *The Shopping Experience* (pp. 92–110). London: Sage.

Bowlby, R. (2003). Commerce and femininity. In D.B. Clarke, M.A. Doel, and K.M.L. Housiaux (Eds), *The Consumption Reader* (pp. 168–172). London: Routledge.

Brennan, Z. (2007). How Tupperware has conquered the world. *Mail Online* (http://www.dailymail.co.uk/femail/article-429672/html, retrieved 27 January 2009).

Browning, L. (2008). Do-it-yourself logos for proud Scion owners. *New York Times*, Monday, 24 March (online edition).

Bryman, A.E. (2004). *The Disneyization of Society*. London: Sage.

de Bury, R. (1909). *The Love of Books: The Philobiblon of Richard de Bury*, trans. E.C. Thomas. London: Chatto & Windus.

Butler, S. (1917). *Erewhon, or, Over the Range*. London: A.C. Fifield.

Cambridge Learner's Dictionary, 3rd edn (2007). Cambridge, United Kingdom: Cambridge University Press.

Campbell, C. (1997). Shopping, pleasure and the sex war. In P. Falk and C. Campbell (Eds), *The Shopping Experience* (pp. 166–176). London: Sage.

Cantwell, A.-M. (2003). Deviant leisure. In J.M. Jenkins and J.J. Pigram (Eds), *Encyclopedia of Leisure and Outdoor Recreation* (p. 114). London: Routledge.

Clarke, D.B., Doel, M.A., and Housiaux, M.L. (2003). General introduction. In D.B. Clarke, M.A. Doel, and K.M.L. Housiaux (Eds), *The Consumption Reader* (pp. 1–23). London: Routledge.

Collins, R. (1981). *Sociology since Midcentury: Essays in Theory Cumulation*. New York: Academic Press.

Cook, D.T. (2006). Leisure and consumption. In C. Rojek, S.M. Shaw, and A.J. Veal (Eds), *A Handbook of Leisure Studies* (pp. 304–316). New York: Palgrave Macmillan.

Critcher, C. (2006). A touch of class. In C. Rojek, S.M. Shaw, and A.J. Veal (Eds), *A Handbook of Leisure Studies* (pp. 271–287). New York: Palgrave Macmillan.

Csikszentmihalyi, M. (1990). *Flow: The Psychology of Optimal Experience*. New York, NY: Harper & Row.

Csikszentmihalyi, M., and Rochberg-Halton, E. (1981). *The Meaning of Things: Domestic Symbols and the Self*. Cambridge, UK: Cambridge University Press.

Curtis, J.E. (1988). Purple recreation. *SPRE Annual on Education*, 3, 73–77.

de Certeau, M. (1984). *The Practice of Everyday Life*, trans. S. Rendall. Berkeley, CA: University of California Press.

de Mille, A. (1952). *Dance to the Piper*. New York: Bantam.

Douglas, M., and Isherwood, B. (1979). *The World of Goods: Towards an Anthropology of Consumption*. New York: Basic Books.

The Economist (2005). Up off the couch. 22 October, p. 35.

The Economist (2008). Play on. 20 December, p. 113.

The Economist (2009). Generation Y goes to work. 3 January, pp. 47–48.

Elgin, D. (1981). *Voluntary Simplicity: Toward a Way of Life that is Outwardly Simple, Inwardly Rich*. New York: William Morrow.

Ershkowitz, H. (2004). Shopping. In G.S. Cross (Ed.), *Encyclopedia of Recreation and Leisure in America*, vol. 2 (pp. 256–260). Detroit, MI: Charles Scribner's Sons.

Etzioni, A. (2003). Introduction: Voluntary simplicity – psychological implications, societal consequences. In D. Doherty and A. Etzioni (Eds), *Voluntary Simplicity: Responding to Consumer Culture* (pp. 1–28). Lanhan, MD: Rowman & Littlefield.

Falk, P., and Campbell, C. (1997). Introduction. In P. Falk and C. Campbell (Eds), *The Shopping Experience* (pp. 1–14). London: Sage.

Fine, G.A. (1979). Small groups and culture creation: The idioculture of Little League baseball teams. *American Sociological Review*, 44, 733–745.

Gans, H.J. (1974). *Popular Culture and High Culture: An Analysis and Evaluation of Taste*. New York: Basic Books.

Gill, B. (1991). The sky line: Disneyitis. *The New Yorker*, 66 (19 April), 96–99.

Glazer, N. (1971). The role of the intellectuals. *Commentary*, 53 (February), 55–61.

Godbey, G. (2004). Contemporary leisure patterns. In G.S. Cross (Ed.), *Encyclopedia of Recreation and Leisure in America*, vol. 1 (pp. 242–248). Detroit, MI: Charles Scribner's Sons.

Godbout, J. (1998). *The World of the Gift*, trans. D. Winkler. Montreal, QC & Kingston, ON: McGill-Queen's University Press.

Goffman, E. (1963). *Stigma: Notes on the Management of Spoiled Identity*. Englewood Cliffs, NJ: Prentice-Hall.

Gouldner, A. (1960). The norm of reciprocity: A preliminary statement. *American Sociological Review*, 25, 161–178.

Hardwick, J. (2008). Sex and the (seventeenth-century) city: A research note towards a long history of leisure. *Leisure Studies*, 27, 459–468.

Harris, D. (2005). *Key Concepts in Leisure Studies*. London: Sage.

Harris, M. (2008). Look-at-me culture wallowing in competitive compassion. *Calgary Herald*, Monday 10 March, p. A6.

Harrison, J. (2001). Thinking about tourists. *International Sociology*, 16, 159–172.

Helft, M. (2008). Tech's late adopters prefer the tried and true. *New York Times*, 12 March (online edition).

Hewer, P., and Campbell, C. (1997). Research on shopping – A brief history and selected literature. In P. Falk and C. Campbell (Eds), *The Shopping Experience* (pp. 186–206). London: Sage.

Holson, L.M. (2008). Hoping to make phone buyers flip. *New York Times*, 29 February (online edition).

Homans, G.C. (1974). *Social Behavior: Its Elementary Forms*, rev. edn. New York: Harcourt, Brace, Jovanovich.

Homer-Dixon, T. (2007). A swiftly melting planet. *The New York Times*, Thursday, 4 October (online edition).

Howard, A. (Ed.) (1995). *The Changing Nature of Work*. San Francisco, CA: Jossey-Bass.

Howe, C. (1995). Factors impacting leisure in middle-aged adults throughout the world: United States. *World Leisure & Recreation*, 37 (1), 37–38.

Jenkins, C., and Sherman, B. (1981). *The Leisure Shock*. London, Eng.: Eyre Methuen.

Johansen, D.O. (1967). *Empire of the Columbia: A History of the Pacific Northwest*, 2nd edn. New York: Harper & Row.

Johnston, J. (2008). The citizen–consumer hybrid: Ideological tensions and the case of Whole Foods Market. *Theory and Society*, 37, 229–270.

Kane, M.J., and Zink, R. (2004). Package adventure tours: Markers in serious leisure. *Leisure Studies*, 23, 329–346.

Katz, J. (1988). *Seductions of Crime: Moral and Sensual Attractions of Doing Evil*. New York: Basic Books.

Keen, A. (2007). *The Cult of the Amateur: How Today's Internet is Killing Our Culture*. New York: Doubleday/Currency.

Kellner, D. (2003). Jean Baudrillard. *Stanford Encyclopedia of Philosophy* (Winter Edition), E.N. Zalta (Ed.). (http://plato.stanford.edu/entries/baudrillard, retrieved April 2008).

Kelly, J.R. (1990). *Leisure*, 2nd edn. Englewood Cliffs, NJ: Prentice-Hall.

Kelly, J.R., and Godbey. G. (1992). *The Sociology of Leisure*. State College, PA: Venture Publishing.

Kiewa, J. (2003). Consumption. In J.M. Jenkins, and J.J. Pigram (Eds), *Encyclopedia of Leisure and Outdoor Recreation* (pp. 79–91). New York: Routledge.

Kodas, M. (2008). *High Crime: The Fate of Everest in an Age of Greed*. New York: Hyperion.

Kohm, S., and Selwood, J. (1998). The virtual tourist and sex in cyberspace. In M. Oppermann (Ed.), *Sex Tourism and Prostitution: Aspects of Leisure, Recreation, and Work* (pp. 123–131). Sydney, Australia: Cognizant Communication Corporation.

Kranz, P. (2008). Measuring wealth by the foot. *New York Times*, 16 March (online edition).

Lambert, R.D. (1996). Doing family history. *Families*, 35, 11–25.

Larrabee, E., and Meyerson, R.B. (Eds) (1958). *Mass Leisure*. Glencoe, IL: Free Press.

Layard, R. (2005). *Happiness: Lessons from a New Science*. New York: Penguin.

Lefevbre, H. (1991). *Critique of Everyday Life, vol. 1, Introduction*, trans. J. Moore. London: Verso.

Lefkowitz, B. (1979). *Breaktime*. New York: Penguin.

Lewis, P. (1995). The urban invasion of rural America: The emergence of the galactic city. In E. Castle (Ed.), *The Changing American Countryside: Rural People and Places* (pp. 39–62). Lawrence, KA: University of Kansas Press.

Lury, C. (1996). *Consumer Culture*. New Brunswick, NJ: Rutgers University Press.

Mackellar, J. (2006). Fanatics, fans, or just good fun? Travel behaviours and motivations of the fanatic. *Journal of Vacation Marketing*, 12, 195–217.

Maffesoli, M. (1996). *The Time of the Tribes: The Decline of Individualism*, trans. D. Smith. London, UK: Sage Publications.

Magliozzi, T., and Magliozzi, R. (2008). Exec needs car that screams success, status. *Calgary Herald*, Thursday, 27 March, NA7.

Maniates, M. (2002). In search of consumptive resistance: The voluntary simplicity movement. In T. Princen, M. Maniates, and K. Conca (Eds), *Confronting Consumption* (pp. 199–236). Cambridge, MA: The MIT Press.

Marcuse, H. (1964). *One-Dimensional Man: Studies in the Ideology of Advanced Industrial Society*. Boston: Beacon Press.

Martin, B., and Mason, S. (1987). Current trends in leisure. *Leisure Studies*, 6, 93–97.

Marx, K. (1977). Grundrisse. In D. McLellan (Ed.), *Karl Marx: Selected Writings*. London, United Kingdom: Oxford University Press.

Mauss, M. (1990). *The Gift: Forms and Function of Exchange in Archaic Societies*. New York: Routledge (original published in French in 1954).

McCall, G.J., and Simmons, J.L. (1978). *Identities and Interactions: An Examination of Human Associations in Everyday Life*, rev. edn. New York: Free Press.

McDonald, M, Wearing, S., and Ponting, J. (2007). Narcissism and neo-liberalism: Work, leisure, and alienation in an era of consumption. *Loisir et Société/Society and Leisure*, 30, 489–510.

Miller, C.C. (2008). For craft sales, the recession is a help. *New York Times*, 23 December (online edition).

Morris, M. (1993). Things to do with shopping centres. In S. During (Ed.), *The Cultural Studies Reader* (pp. 391–409). London: Routledge.

Moubarac, J.-C., Gupta, R., and Martin, I. (2007). La promotion du poker sur Internet et son influence sur la participation des jeunes adultes aux jeux d'argent. *Loisir et Société/Society and Leisure*, 30, 513–525.

Nava, M. (1992). *Changing Cultures: Feminism, Youth and Consumerism*. London: Sage.

Nava, M. (1997). Women, the city and the department store. In P. Falk and C. Campbell (Eds), *The Shopping Experience* (pp. 56–91). London: Sage.

Nazareth, L. (2007). *The Leisure Economy: How Changing Demographics, Economics, and Generational Attitudes Will Reshape Our Lives and Our Industries*. Mississauga, ON: John Wiley & Sons Canada.

Norwegian Ministry of the Environment (1994). *Oslo Roundtable on Sustainable Production and Consumption*. Oslo, Norway: Government of Norway.

Oppermann, S., McKinley, S., and Chon, K.-S. (1998). Marketing sex and tourism destinations. In M. Oppermann (Ed.), *Sex Tourism and Prostitution: Aspects of Leisure, Recreation, and Work* (pp. 20–29). Sydney, Australia: Cognizant Communication Corporation.

Oswald, A.J., and Powdthavee, N. (2007). Review 1: Obesity, unhappiness, and the challenge of affluence: Theory and evidence, *Economic Journal*, 117 (521), F441–454.

Parker, S. (1983). *Leisure and Work*. London: George Allen & Unwin.

Pearce, J.L. (1993). *Volunteers: The Organizational Behavior of Unpaid Workers*. London, UK: Routledge.

Plog, S. C. (1991). *Leisure Travel: Making It a Growth Market…Again!* New York: Wiley.

Prus, R., and Dawson, L. (1991). Shop 'til you drop: Shopping as recreational and laborious activity. *Canadian Journal of Sociology*, 16, 145–164.

Putnam, R.D. (2000). *Bowling Alone: The Collapse and Revival of American Community*. New York: Simon & Schuster.

Pynn, L. (2008). 1,300 birds killed in B.C. gambling busts. *Calgary Herald*, Saturday, 1 March, p. A13.

Rifkin, J. (1995). *The End of Work: The Decline of the Global Labor Force and the Dawn of the Post-Market Era*. New York, NY: G.P. Putnam's Sons.

Ritzer, G. (1993). *The McDonaldization of Society: An Investigation into the Changing Character of Contemporary Social Life*. Thousand Oaks, CA: Pine Forge Press.

Roberts, K. (1999). *Leisure in Contemporary Society*. Wallingford, Oxon: CABI Publishing.

Robinson, J.P., and Godbey G. (1997). *Time for Life: The Surprising Ways Americans Use Their Time*. University Park, PA: Pennsylvania State University Press.

Robinson, J.P., and Martin, S. (2008). What do happy people do? *Social Indicators Research*, 89, 565–571.

Robinson, T. (2008). Explaining the vicious cycle of overwork and overconsumption. In J. Caudwell, S. Redhead, and A. Tomlinson (Eds), *Relocating the Leisure Society: Media, Consumption and Space* (LSA Publication No. 101) (pp. 55–66). Eastbourne, United Kingdom: Leisure Studies Association.

Rocher, G. (1972). *A General Introduction to Sociology: A Theoretical Perspective*, trans. by P. Sheriff. Toronto, ON: Macmillan Co. of Canada.

Rojek, C. (1993). Disney culture. *Leisure Studies*, 12, 121–136.
Rojek, C. (1997). Leisure theory: Retrospect and prospect. *Loisir et Société/Society and Leisure*, 20, 383–400.
Rojek, C. (2000). *Leisure and Culture*. London: Palgrave Macmillan.
Rojek, C. (2005). *Leisure Theory: Principles and Practice*. London: Palgrave Macmillan.
Rosenberg, B., and White, D.M. (1957). *Mass Culture: The Popular Arts in America*. Glencoe, IL: Free Press.
Samuel, N. (1994). *New Routes for Leisure* (pp. 45–57). Lisbon, Portugal: Instituto de Ciências Sociais, Universidade de Lisboa.
Scammell, M. (2000). The Internet and civic engagement: The age of the citizen–consumer. *Political Communication*, 17, 351–355.
Schickel, R. (1986). *The Disney Version: The Life, Times, Art, and Commerce of Walt Disney*, rev. edn. London: Pavilion.
Schor, J.B. (1991). *The Overworked American: The Unexpected Decline of Leisure*. New York: Basic Books.
Seligman, M.E.P. (2003). *Authentic Happiness: Using the New Positive Psychology to Realize Your Potential for Lasting Fulfillment*. New York: Free Press.
Shaya, G. (2004). The *flâneur*, the *badaud*, and the making of a mass public in France, circa 1860–1910. *The American Historical Review*, 109 (1), 41–78.
Sherman, B. (1986). *Working at Leisure*. London, Eng.: Methuen London.
Shields, R. (1966). Foreword: Masses or tribes? In M. Maffesoli. *The Time of the Tribes: The Decline of Individualism*, trans. D. Smith (pp. ix–xii). London, UK: Sage Publications.
Siegenthaler, K.L., and O'Dell, I. (2003). Older golfers: Serious leisure and successful aging. *World Leisure Journal*, 45 (1), 45–52.
Silverberg, S. (2008). Employee perceptions of organizational commitment: An exploratory study. Ph.D. Dissertation, Department of Sociology, University of Calgary.
Simmel, G. (1971). Fashion. In D.N. Levine (Ed.), *George Simmel on Individuality and Social Forms*. Chicago, IL: University of Chicago Press.
Smith, D.H. (2000). *Grassroots Associations*. Thousand Oaks, CA: Sage Publications.
Smith, D.H., Stebbins, R.A., and M. Dover (2006). *A Dictionary of Nonprofit Terms and Concepts*. Bloomington, IN: Indiana University Press.
Snyder, C.R., and Lopez, J. (2007). *Positive Psychology: The Scientific and Practical Explorations of Human Strengths*. Thousand Oaks, CA: Sage.
Spaargaren, G. (2005). Sustainable consumption: A theoretical and environmental policy. In D. Southerton, H. Chappells, and B. van Vliet (Eds), *Sustainable Consumption: The Implications of Changing Infrastructures of Provision* (pp. 15–31). Cheltenham, United Kingdom: Edward Elgar.
Stalp, M.C. (2007). *Quilting: The Fabric of Everyday Life*. New York: Berg.
Stebbins, R.A. (1979). *Amateurs: On the Margin Between Work and Leisure*. Beverly Hills, CA: Sage (also available at www.soci.ucalgary.ca/seriousleisure – Digital Library).
Stebbins, R.A. (1981). The social psychology of selfishness. *Canadian Review of Sociology and Anthropology*, 18, 82–92.
Stebbins, R.A. (1982). Serious leisure: A conceptual statement. *Pacific Sociological Review*, 25, 251–272.

Stebbins, R.A. (1992). *Amateurs, Professionals, and Serious Leisure*. Montreal, QC & Kingston, ON: McGill-Queen's University Press.

Stebbins, R.A. (1994). The liberal arts hobbies: A neglected subtype of serious leisure. *Loisir et Société/Society and Leisure*, 16, 173–186.

Stebbins, R.A. (1995). Leisure and selfishness: An exploration. In G. S. Fain (Ed.), *Reflections on the Philosophy of Leisure, Vol. II, Leisure and Ethics* (pp. 292–303). Reston, VA: American Alliance for Health, Physical Education, Recreation, and Dance.

Stebbins, R.A. (1996a). Volunteering: A serious leisure perspective. *Nonprofit and Voluntary Action Quarterly*, 25, 211–224.

Stebbins, R.A. (1996b). *The Barbershop Singer: Inside the Social World of a Musical Hobby*. Toronto, ON: University of Toronto Press.

Stebbins, R.A. (1996c). *Tolerable Differences: Living with Deviance* (2nd edn). Toronto, ON: McGraw-Hill Ryerson (also available at www.soci.ucalgary.ca/ seriousleisure – Digital Library).

Stebbins, R.A. (1996d). Cultural tourism as serious leisure. *Annals of Tourism Research*, 23, 948–950.

Stebbins, R.A. (1997a). Casual leisure: A conceptual statement. *Leisure Studies*, 16, 17–25.

Stebbins, R.A. (1997b). Identity and cultural tourism. *Annals of Tourism Research*, 24, 450–452.

Stebbins, R.A. (1998). *The Urban Francophone Volunteer: Searching for Personal Meaning and Community Growth in a Linguistic Minority*. Vol. 3, No. 2 (New Scholars-New Visions in Canadian Studies quarterly monographs series). Seattle, WA: University of Washington, Canadian Studies Centre.

Stebbins, R.A. (2000). Obligation as an aspect of leisure experience. *Journal of Leisure Research*, 32, 152–155.

Stebbins, R.A. (2001a). *New Directions in the Theory and Research of Serious Leisure*, Mellen Studies in Sociology, vol. 28. Lewiston, NY: Edwin Mellen.

Stebbins, R.A. (2001b). Volunteering – mainstream and marginal: Preserving the leisure experience. In M. Graham and M. Foley (Eds), *Volunteering in Leisure: Marginal or Inclusive?* (Vol. 75, pp. 1–10). Eastbourne, United Kingdom: Leisure Studies Association (also available at www.soci.ucalgary.ca/seriousleisure – Digital Library, Other Works).

Stebbins, R.A. (2001c). The costs and benefits of hedonism: Some consequences of taking casual leisure seriously. *Leisure Studies*, 20, 305–309.

Stebbins, R.A. (2002). *The Organizational Basis of Leisure Participation: A Motivational Exploration*. State College, PA: Venture Publishing.

Stebbins, R.A. (2004a). *Between Work and Leisure: The Common Ground of Two Separate Worlds*. New Brunswick, NJ: Transaction Publishers.

Stebbins, R.A. (2004b). Pleasurable aerobic activity: A type of casual leisure with salubrious implications. *World Leisure Journal*, 46(4), 55–58 (also available at www.soci.ucalgary.ca/seriousleisure – Digital Library, Other Works).

Stebbins, R.A. (2005a). Choice and experiential definitions of leisure. *Leisure Sciences*, 27, 349–352.

Stebbins, R.A. (2005b). *Challenging Mountain Nature: Risk, Motive, and Lifestyle in Three Hobbyist Sports*. Calgary, AB: Detselig.

Stebbins, R.A. (2005c). Project-based leisure: Theoretical neglect of a common use of free time.*Leisure Studies*, 24, 1–11.

Stebbins, R.A. (2005d). The role of leisure in arts administration. *Occasional Paper Series*, Paper No. 1. Eugene, OR: Center for Community Arts and Public Policy, University of Oregon Arts. (published online at: http://aad.uoregon.edu/icas/documents/stebbins0305.pdf).

Stebbins, R.A. (2006). Shopping as leisure, obligation, and community. *Leisure/Loisir*, 30, 475–486.

Stebbins, R.A. (2007a). *Serious Leisure: A Perspective for Our Time*. New Brunswick, NJ: Transaction Publishers.

Stebbins, R.A. (2007b). A leisure-based, theoretic typology of volunteers and volunteering. *Leisure Studies Association Newsletter*, 78 (November), 9–12 (also available at www.soci.ucalgary.ca/seriousleisure – Digital Library, 'Leisure Reflections No.16').

Stebbins, R.A. (2007c). Leisure's role in voluntary simplicity. *Leisure Studies Association Newsletter*, no. 77 (July), 16–20. (also available at www.soci.ucalgary.ca/seriousleisure – Digital Library, 'Leisure Reflections No. 15').

Stebbins, R.A. (2007d). Leisure studies: The happy science. *Leisure Studies Association Newsletter*, 76 (March), 20–22. (also available at www.soci.ucalgary.ca/seriousleisure – Digital Library, 'Leisure Reflections No. 14').

Stebbins, R.A. (2009a). New leisure and leisure customization. *World Leisure Journal*, 51, in press.

Stebbins, R.A. (2009b). *Personal Decisions in the Public Square: Beyond Problem Solving into a Positive Sociology*. New Brunswick, NJ: Transaction Publishers.

Stebbins, R.A. (2009c). Nature challenge activities: Serious leisure in natural settings. *Leisure Studies Association Newsletter*, no. 84 (November), in press. (also available at www.soci.ucalgary.ca/seriousleisure – Digital Library, 'Leisure Reflections No. 22').

Stebbins, R.A. (in press). Les Frontières entre les loisirs et le travail: Cinq ponts. In C. Marry, A. Degenne, and S. Moulin (Eds), *Les catégories sociales et leurs frontières*.

Stone, G.P. (1954). City shoppers and urban identification: Observations on the social psychology of city life. *American Journal of Sociology*, 60, 36–45.

Stone, G.P. (1962). Appearance and the self. In A.M. Rose (Ed.), *Human Behavior and Social Processes: An Interactionist Approach* (pp. 86–118). Boston, MA: Houghton Mifflin.

Storr, M. (2003). *Latex and Lingerie: Shopping for Pleasure at Ann Summers Parties*. Oxford, UK: Berg.

Stuever, H. (2008). Selling to the anti-consumer. *Calgary Herald*, Monday, 31 March, p. B9.

Swartzberg, J. (2007). Bottled water bites the dust. *University of California, Berkeley, Wellness Letter*, 23 (12), 3.

Tanner, J., Asbridge, M., and Wortley, S. (2008). Our favourite melodies: musical consumption and teenage lifestyles. *British Journal of Sociology*, 59, 117–144.

Taylor, C. (1991). *The Malaise of Modernity*. Toronto, ON: The House of Anancy Press.

Taylor-Gooby, P. (2006). The rise (or not) of the citizen–consumer. ESRC Society Today (www.regard.ac.uk, retrieved 17 November 2006).

Toffler, A. (1980). *The Third Wave: The Classic Study of Tomorrow*. New York: William Morrow.

Trebay, G. (2008). Luxury prices are falling; the sky, too. *New York Times*, Thursday 4 December (online edition).

Truzzi, M. (1972). The occult revival as popular culture. *The Sociological Quarterly*, 13, 16–36.

Turner, R.H., and Killian, L.M. (1987). *Collective Behavior*, 3rd edn. Englewood Cliffs, NJ: Prentice-Hall.

United Nations Environment Programme (UNEP) and World Tourism Organization (WTO) (2005). *Making Tourism More Sustainable: A Guide for Policymakers*. Paris & Madrid: UNEP and WTO.

Unruh, D.R. (1979). Characteristics and types of participation in social worlds. *Symbolic Interaction*, 2, 115–130.

Unruh, D.R. (1980). The nature of social worlds. *Pacific Sociological Review*, 23, 271–296.

Urry, J. (1994). Cultural change and contemporary tourism. *Leisure Studies*, 13, 233–238.

VandeSchoot, L. (2005). Navigating the divide: Muslim perspectives on Western conceptualizations of leisure. Masters Thesis, Wageningen University, Social Spatial Analysis Chair Group.

Veblen, T. (1899). *The Theory of the Leisure Class: An Economic Study of Institutions*. New York: Macmillan.

Warde, A. (2005). Consumption. In T. Bennett, L. Grossberg, and M. Morris (Eds), *New Keywords: A Revised Vocabulary of Culture and Society* (pp. 57–59). Oxford, UK: Blackwell.

Wearing, S.L. (2001). *Volunteer Tourism: Seeking Experiences that Make a Difference*. Wallingford, Oxon, UK: CAB International.

Wilensky, R. (1978). A conceptual analysis of the verbs *need* and *want*. *Cognitive Science*, 2, 391–396.

Williams, R.M., Jr. (2000). American society. In E.F. Borgatta, and R.J.V. Montgomery (Eds), *Encyclopedia of Sociology*, 2nd edn, Vol. 1 (pp. 140–148). New York: Macmillan.

Winter, M. (2007). *Sustainable Living: For Home, Neighborhood and Community*. Napa, CA: Westsong Publishing.

Wolff, J. (1985). The invisible *flâneuse*: Women in the literature of modernity. *Theory, Culture & Society*, 2(3), 37–46.

Woodhouse, A. (1989). *Fantastic Women: Sex, Gender, and Transvestitism*. New Brunswick, NJ: Rutgers University Press.

Yankelovich, D. (1981). *New Rules: Searching for Self-Fulfillment in a World Turned Upside Down*. New York, NY: Random House.

Zukin, S. (1996). *The Culture of Cities*. Malden, MA: Blackwell.

Zukin, S., and Maguire, J.S. (2004). Consumers and consumption. *Annual Review of Sociology*, 30, 173–197.

Zuzanek, J. (1996). Canada. In G. Cushman, A.J. Veal, and J. Zuzanek (Eds), *World Leisure Participation: Free Time in the Global Village* (pp. 35–76). Wallingford, Oxon, England: CAB International.

Index

Aberle, D.F., 149–50
amateurs, *see* serious leisure,
    amateurs and
apparel, 48–9
Arai, S.M., 19
Aronowitz, S., 77
Asbridge, M., 45

Back, K.W., 136
Barber, B.R., 97
Baudrillard, J., 60–3, 96, 98
Bauman, Z., 7, 68–9
Belk, R., 3–4, 75, 108
Bell, D., 6
Benjamin, W., 91
Bennett, A., 52
Binham, C., 44
Blackshaw, T., 139
Booth, W., 140
Bott, E., 139
Bourdieu, P., 65–8, 73
Bowen, C.D., 127
Bowlby, R., 4, 88, 90–1
Brennan, Z., 100
Britt, R., 156–7
Browning, L., 46
Bryman, A.E., 72–3
Buchsbaum, J., 79
Butler, S., 128–9

Campbell, C., 4, 88, 89, 90, 161
Cantwell, A.-M., 27
casual leisure, 22–4
    conspicuous consumption, and,
        43–4
    definition of, 22–3
    as deviant, 26–7
    Disneyfication leading to, 73
    initiatory consumption in, 110
    nonconsumptive, 118–20
    reputation of, 81
    as shopping, 91–2, 94
    and simple living, 153

as tourism, 47–8
in tribes, 50–1, 146
volunteers and, 148
and waste, 53–4
see also organization of consumptive
    leisure
Chon, K.-S., 105
Clarke, D.B., 1, 85, 160
clothing, *see* apparel
Collins, R., 67
competitive compassion, 40–2
conspicuous consumption, 30–55,
    110, 132
    adoption stages and, 40
    gifts in, 35–8
    identity and, 44–52, 97–8
    invidious comparison, 31
    kula as, 34–5, 110
    in leisure, 42–4
    and the leisure class, 31–2, 54
    potlatch as, 33–4, 35, 36, 44, 110
    primitive, 33–5
    reciprocity and, 37–8, 110
    through apparel, 48–9
    through technology, 38–40
    through tourism, 46–8
    today, 32–3, 38–44, 53–4, 54–5, 62–3
    tribes and, 49–52
    Veblen's theory of, 30–3
    and waste, 52–4
    see also competitive compassion;
        new goods and services
conspicuous waste, 52–4
consumers, 3–4
    agency of, 61–2, 64–5, 80, 135
    as buffs, 43, 146
    as citizen-consumers, 101–3, 105
    as fans, 43, 146, 160
    knowledge (information) of, 74–5,
        89, 93, 102–3, 104, 118, 137,
        138–9, 139–40, 141, 142, 144
    needs, 96–7

as playful (inventive), 61–2,
64–5, 80
as prosumers, 98–9, 105
wants and, 90–1, 96–8, 105, 159
see also conspicuous consumption;
consumption; shopping
consumer society, 6–7, 60, 62
consumption, 3–4
in the 19ᵗʰ century, 56–7
adoption stages and, 39–40, 53,
97–8, 137–8, 152
anthropology of, 73–5
authenticity and, 130–1
Baudrillard on, 60–3
Bauman on, 7, 68–9
confidence and, 131–2
conspicuous, 30–55
de Certeau on, 63–5
definition of, 3–4, 74, 75, 160
deviant, 103–6
Disneyfication and, 73
Disneyization and, 71–3
Douglas on, 74–5
economists' view of, 132
and happiness, 128–30
as having, 108, 130, 131
Isherwood on, 74–5
Lefevbre on, 59–60
and leisure, 2–3, 7, 54–5, 82–107,
97–8, 108–33, 159
and leisure customization, 7–9
McDonaldization (Ritzer) and,
69–71
Marx on, 56–7
mass, 4, 5–6, 8
as monetary acquisition, 4
obligatory, 82–3, 87–8, 97–8, 103,
109, 110, 111, 143
positive side of, 59, 64–5, 67, 73–5,
78–9, 80, 94, 128–30, 158–9
as practical, 132
Simmel on, 58–9
and social class, 85–6
and social movements, 149–58
through leisure service
organizations, 143
two-phase model of, 4, 82–107,
108–33

see also competitive compassion;
consumers; facilitative
consumption; initiatory
consumption; new goods and
services; organization of
consumptive leisure; shopping;
simple living; sustainable
consumption
consumption studies, 1–3, 56, 82,
106–7
macrosociological nature of, 28, 56
Cook, D.T., 2
Critcher, C., 7
Csikszentmihalyi, M., 21–2, 43
Curtis, J.E., 104

Dawson, L., 4, 87–8
de Bury, R., 116
de Certeau, M., 63–5, 98
de Mille, A., 115–16
DiFazio, W., 77
direct sales parties, *see* shopping
Disneyfication, 73
Disneyization, 71–3
compared with McDonaldization,
73
Doel, M.A., 1, 85, 160
Douglas, M., 74–5, 159
Dover, M., 15
Dunning, L., 119–20

Elgin, D., 150–1
Ershkowitz, H., 85–6
Etzioni, A., 130

facilitative consumption, 42–3,
111–18, 131
definition of, 111
Falk, P., 4, 88
Farnish, C., 102
fashion, 58–9
as initiatory consumption, 59
flâneur, 91–2, 161
Franz, P., 33

Gandhi, M., 150, 152
Gans, H., 5, 6
gender, *see* shopping, and gender
gift, 4, 35–8, 60, 74

Gill, B., 71
Glazer, N., 6
Godbey, G., 7–9, 76, 98, 137
Godbout, J., 35, 36
Goffman, E., 62, 128
Gouldner, A., 37
Gregg, R., 151, 152
Gupta, R., 104, 161

Hardwick, J., 104
Harris, D., 71, 160
Harris, M., 40
Harrison, J., 12
Helft, M., 39–40
Hewer, P., 90
hobbyists, *see* serious leisure, hobbyists and
Holson, L.M., 98–9
Homans, G.C., 37
Homer-Dixon, T., 148
Housiaux, M.L., 1, 85, 160
Howard, A., 77
Howe, C., 76

identity, 44–52, 112
    apparel and, 48–9
    authenticity and, 131
    symbols, 45–6
    tourism and, 46–8
    in tribes, 145–6
    types of, 5, 44
    wants and, 97–8
initiatory consumption, 42, 59, 109–11
    definition of, 109
Isherwood, B., 74–5, 159

Jenkins, C., 160
Johansen, D.O., 34
Johnston, J., 101–2

Kane, M.J., 47–8
Katz, J., 27
Keen, A., 23
Kellner, D., 61
Kelly, J.R., 10, 137
Kiewa, J., 2, 118
Killian, L.M., 148

Kodas, M., 43
Kohm, S., 105

Lambert, R.D., 117
Larrabee, E., 5
Layard, R., 129–30
Lee, K.J., 114
Lefevbre, H., 59–60, 97, 98
Lefkowitz, B., 76
leisure, 9–13
    as activity, 10–13, 87, 154, 160
    Baudrillard on, 61, 63
    boredom and, 10
    Bourdieu on, 65–8
    Bryman on, 72–3
    constraints on, 10
    and consumption, 2–3, 7, 82–107, 108–33, 159
    as core activity, 11–13, 115, 120, 127, 132
    definition of, 10
    deviant, 27–8, 103–6, 159, 160
    as doing, 108, 130, 131
    economics of, 78–9
    generational analysis of, 78–9
    the happy science and, 133, 158, 159
    incomplete conceptualization of, 80–1
    Lefevbre on, 60
    nonconsumptive, 117, 118–26
    opportunities for, 77
    Ritzer on, 70
    as role, 11, 160
    and selfishness, 127–8, 159
    and social class, 7, 54
    in social movements, 147, 148, 149–58
    social worlds of, 144, 146
    time, 75–6, 78, 81, 127, 152
    as uncontrollable, 127–8, 159
    see also organization of consumptive leisure; serious leisure perspective; simple living; sustainable consumption
leisure customization, 7–9, 98
lifestyle, 50, 55, 150–3, 154, 155
Long, J., 139

Lopez, S., 130
Lury, C., 7

McCall, G.J., 44
McDonaldization, 69–71, 73
    compared with Disneyization, 73
McDonald, M., 2–3
MacKellar, J., 128
McKinley, S., 105
Maffesoli, M., 49–51, 145, 146
Magliozzi, R., 32–3
Magliozzi, T., 32–3
Maguire, J.S., 1–2
Maniates, M., 154–5
Marcuse, H., 6, 63
Martin, B., 92
Martin, I., 104, 161
Martin, S., 88–9
Marx, K., 56–7, 59, 60–3, 70
Mason, S., 92
mass consumption (leisure/culture), 4,
        5–6, 8, 145
    critique of, 5–6, 60, 63, 68–9, 71–3,
        110–11
    history of, 85–6
Mauss, 34–5, 36, 37, 66
Meyerson, R.B., 5
Miller, C.C., 116–17
Moubarac, J.-C., 104, 161

Nava, M., 89, 90
Nazareth, L., 78–9, 159
new goods and services, 43
    adoption stages of, 39–40
    non-adoption of, 40, 53
nonconsumptive leisure, 117, 118–26
    as casual leisure, 118–20
    definition of, 118
    as serious leisure, 120–4, 161

obligation, 95, 147, 148, 152, 161
occupational devotion, 112–15
O'Dell, I., 19
Oppermann, S., 105
organization of consumptive leisure,
        134–59
    agency in, 135
    in consumer-related social
        movements, 149–58
    definition of, 134–6
    in grassroots associations, 140–1
    in leisure service organizations,
        142–3
    sharing knowledge/information
        through, 137, 138–9, 139–40,
        141, 142, 144, 161
    in small groups, 136–9, 161
    social appeal of, 137
    in social movements, 147–58
    in social networks, 139–40
    in social worlds, 143–5, 146
    and solitary leisure, 135
    in tribes, 51–2, 145–7
    in voluntary organizations, 141–2,
        158, 161
Oswald, A.J., 129

Parker, S., 10
Pearce, J.L., 142
Pedlar, A.M., 19
Plog, S.C., 47
Ponting, J., 2–3
positiveness, *see* consumption;
        positive sociology
positive sociology, 9–10
    definition of, 9–10
    and happiness, 129
    and volunteering, 16
Powdthavee, N., 129
project-based leisure, 24–6
    as bridging initiatory/facilitative
        consumption, 118
    definition of, 24
    facilitative consumption in, 117–18
    initiatory consumption in, 110
    shopping as, 95–6
    and simple living, 53
    see also organization of consumptive
        leisure
prosumers, 98–9
Prus, R., 4, 87–8
Putnam, R.D., 100
Pynn, L., 8

reciprocity, 37–8
Rifkin, J., 76–7, 160
Ritzer, G., 69–71, 72
Roberts, K., 2, 51, 118

Robinson, J.P., 76, 88–9
Robinson, T., 79
Rochberg-Halton, E., 43
Rocher, G., 134–5
Rogers, W., 151, 162
Rojek, C., 23, 27, 52–3, 71
Rosenberg, B., 6

Samuel, N., 76
Scammel, M., 101
Schickel, R., 73
Schor, J., 75–6
Seligman, M.E.P., 129
Selwood, J., 105
serious leisure, 14–22
    amateurs and, 15, 115, 124
    conspicuous consumption, and,
        43, 47
    definition of, 14–15
    as deviant, 26–7
    devotees/participants in, 19, 139–40
    facilitative consumption in, 115–17,
        143
    as fanatical, 128
    flow and, 21–2
    frequency of, 162
    hobbyists and, 15, 99, 116–17,
        120–4, 157–8, 161
    initiatory consumption in, 110
    Lefevbre on, 60
    as marginal, 128
    motivation in, 19–22
    nonconsumptive, 120–4
    qualities of, 17–19
    rewards of, 19–21
    and selfishness, 127–8
    and simple living, 152, 153
    as tourism, 47
    in tribes, 50–1, 146
    as uncontrollable, 127–8
    volunteers and, 15–17, 120, 142,
        148, 161
    and waste, 53–4
    as work, 112–15
    see also organization of consumptive
        leisure

serious leisure perspective, 13–26, 111,
        160
    and agency, 67
    casual leisure component of, 22–4
    contextual component of, 28–9,
        56–81
    definition of, 13
    project-based leisure component of,
        24–6
    serious leisure component of, 14–22
Shaya, G., 91–2
Sherman, B., 160
Shields, R., 50
shopping, 83–96, 99–103, 159
    as activity, 87
    agency in, 83–4
    and citizen-consumers, 101–3
    and community, 99–103
    context of, 94–6, 105
    distance, 93–4, 104–5
    and gender, 89–91, 100, 161
    history of, 84–6
    knowledge and, 89, 93, 102–3, 104
    as leisure, 87–9, 90, 91–6
    nature of, 83–4
    as obligation, 86–9, 90, 92, 94, 95
    organizational influence on, 138
    and serious leisure, 92–3
    shopaholics and, 96
    and social capital, 100–1, 105–6
    through direct selling, 99–100,
        105–6
    types of, 87–9
    window, 91–2, 93
    see also tourism, shopping and
Siegenthaler, K.L., 19
Silverberg, S., 130–1
Simmel, G., 58–9
Simmons, J.L., 44
simple living, 53–4, 55, 150–3
    definition of, 150–1
    history of, 150–1, 152–3
    leisure and, 152, 153, 160
Smith, D.H., 15, 140–1
Snyder, C.R., 130
social class, *see* conspicuous
        consumption; leisure
Spaargaren, G., 154

Stebbins, R.A., 2, 4, 8, 10–26, 27–8,
    43, 47, 51, 91, 92–3, 95, 103,
    112–26, 133, 134, 136, 143,
    146, 149, 153, 154, 158, 159,
    160, 161, 162
Stone, G.P., 48–9, 87
Storr, M., 100
Stuever, H., 40
sustainable consumption, 153–8,
    159
  definition of, 153–4
  leisure and, 155–8, 162
  and simple living, 154, 155
  social practices model of, 154
Swartzberg, J., 161

Tanner, J., 45
Taylor, C., 130
Taylor-Gooby, P., 102–3
Thoreau, H.D., 152
Toffler, A., 98
tourism, 46–8
  shopping and, 86, 92, 105
  and sustainable consumption,
    157
Trebay, G., 49
tribes (global postmodern), 49–52,
    145–7
Truzzi, M., 28
Turner, R.H., 148
two-phase model of consumption, 4,
    82–107, 108–33

Unruh, D., 18–19, 143–4
Urry, J., 46–7

VandeSchoot, L., 8
Veblen, T., 30–55, 62–3, 67, 132
voluntary simplicity, *see* simple living
volunteers, *see* serious leisure,
    volunteers

Warde, A., 61
waste, *see* conspicuous waste
Wearing, S.L., 2–3, 120
Weber, M., 69, 70
White, D.M., 6
Wilensky, R., 96
Williams, R.M., Jr, 112
Winter, M., 155
Wolff, J., 161
Woodhouse, A., 105
work, 76–7
  facilitative consumption at, 111–12,
    112–15
  and leisure, 78–9
  occupational devotion as, 112–15
  reduction of, 76–7
Wortley, S., 45

Yankelovich, D., 76

Zink, R., 47–8
Zukin, S., 1–2, 71
Zuzanek, J., 76